# An Illustrated
# Guide to
# ARIZONA WEEDS

# An Illustrated
# Guide to
# ARIZONA WEEDS

**Kittie F. Parker**

*Drawings by*
*Lucretia Breazeale Hamilton*

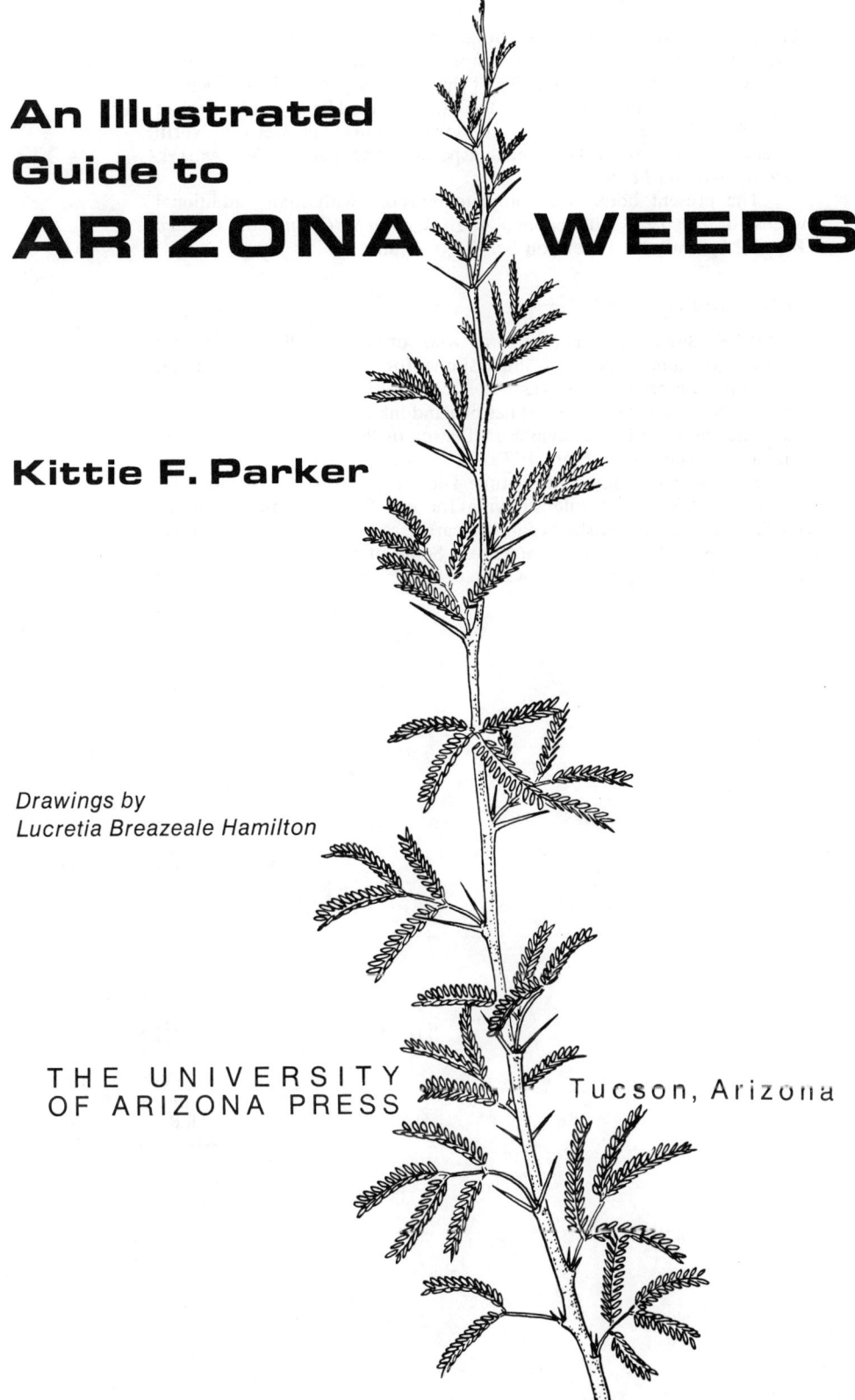

THE UNIVERSITY
OF ARIZONA PRESS

Tucson, Arizona

## About the Author —

KITTIE F. PARKER, Professor of Botany at George Washington University, and Research Associate at the Smithsonian Institution, Washington, D.C., was formerly on the faculty of the University of Arizona Department of Botany, and Curator of the Herbarium. Her primary field of research is in the Compositae (sunflower family) of western North America, in the tribe Helenieae, especially the genera *Hymenoxys, Tetraneuris,* and *Pectis.*

The present book is a complete revision, with many additional weeds, of the 1958 publication, *Arizona Ranch, Farm, and Garden Weeds,* Agricultural Extension Service, Circular 265.

## About the Illustrator —

LUCRETIA BREAZEALE HAMILTON, who drew the illustrations for *Weeds,* has done botanical illustrations for numerous publications. The Soil Conservation Service, the University of Arizona, and the Forestry Service have employed her pen and ink drawings. She illustrated the cactus publications of Lyman D. Benson of Pomona College, and his chapter on cacti in *The Flora of Texas.*

She and her family are long time Tucson residents, who came to the state from Virginia. While studying for her BS in botany from the University of Arizona, she began her career as botanical artist by doing illustrations for her botany notebooks. Some of her first work was for E. D. Ball, the entomologist, who taught her to draw grasshoppers in pen and ink.

*Fourth Printing 1990*

THE UNIVERSITY OF ARIZONA PRESS

I. S. B. N. 0–8165–0288–9
L. C. No. 72–75471

## FOREWORD

Weed identification is the first battle in the endless struggle with weeds, and Kittie Parker in this book has assembled the material for effective combat. In the 1970s, problems must be faced which did not even exist at the time of her earlier Agricultural Bulletin, *Arizona Ranch, Farm, and Garden Weeds.* Weed populations slowly change, so new weeds have to be discussed. Land use in Arizona has changed, and agricultural weeds — such as pigweed and Johnsongrass — have been replaced by the spurge and crabgrass of urban areas. New chemicals — the herbicides — have been developed to provide an additional alternative when weeds must be controlled.

All these elements plus those presented in the first edition in bulletin form are unfolded in Kittie Parker's proven, effective descriptions and discussions, augmented by the excellent drawings by Lucretia Hamilton. Identification and control features of the first edition established the compilation as of working value not only to those on farms, ranches and in urban areas, but also to science classes in schools, to field men of the chemical industries working on crops, and to our neighbors in northwestern Mexico whose many weed problems are the same because of the similarity of the regions.

This new, updated version will continue to perform such a valuable role.

K. C. HAMILTON
Agronomist, University of Arizona

## ACKNOWLEDGEMENTS

I am especially indebted to Mrs. Lucretia B. Hamilton for her excellent illustrations; to Walter S. Phillips of the University of Arizona Biological Sciences Department and the late Charles U. Pickerell, former Director of the Agriculture Extension Service, for without their assistance, the first edition of *Weeds* could not have been published; to Keith C. Hamilton of the Agronomy Department, for his advice and information on the important farm weeds in Arizona and their distribution; to William J. Pistor, former Head of the Department of Animal Pathology, for information on the livestock-poisoning weeds; and to my husband Kenneth W. Parker, of the U.S. Forest Service, for his considerable aid, and for information on the range weeds. Gratitude is expressed to the curators of the U.S. National Herbarium, and especially to Charles T. Mason, Jr., curator of the University of Arizona Herbarium, for their helpful suggestions and the use of their materials and facilities; and to the many people who helped in the collecting of fresh plant material for the drawings, particularly Louis B. Hamilton. Thanks are also due to the University of Arizona Press for effecting publication.

K. F. P.

# CONTENTS

# GLOSSARY

*Achene* — A small dry one-seeded fruit in which the ovary wall is free from the seed

*Anther* — Enlarged part at the top of the stamen; bears the pollen

*Awn* — A bristle or spine, as at the tip of some bracts of grasses

*Axil* — The upper angle between the stem and the leaf

*Bract* — A modified leaf underneath a flower or inflorescence

*Bur* — A rough or prickly envelope around a seed or fruit

*Calyx* — The outer set of floral leaves, usually green, composed of all the sepals

*Corolla* — The set of floral leaves, usually showy, composed of all the petals

*Deciduous* — Dropping off; shedding of leaves at the end of the growing season

*Dehiscent* — Splitting open at maturity

*Disk flowers* — The tubular flowers in the center of the flower heads in the sunflower family

*Embryo* — A young plant enclosed in a seed

*Floret* — A small flower in the spikelet of the grasses, or the flower head of the sunflower family

*Flower head* — A dense inflorescence of stalkless flowers

*Frond* — The leaf of a fern

*Fruit* — The seed case or seed pod (the ripened ovary); containing anywhere from one to many seeds

*Glands* — Secreting tissue

*Glumes* — A pair of bracts at the base of the grass spikelet

*Grain* — The seedlike dispersal unit of the grass family

*Inferior ovary* — Where the sepals, petals, and stamens appear to rise from the very top of the ovary

*Inflorescence* — The arrangement of flowers on a stalk where several are present, as a spike, raceme, head, panicle

*Involucre* — A cluster of bracts

*Keel* — A central dorsal ridge; the two united petals in some legume flowers

*Leaflet* — A division of a compound leaf

*Mycorrhiza* — Formed by a combination of underground threads and rootlets

*Node* — That position on the stem where a leaf, leaves, or branch arises

*Nutlet* — A small nut; the one-seeded portion of a larger fruit that separates at maturity

*Ovary* — Base of the female part of the flower (pistil); ripens into the fruit

*Panicle* — A loose, irregularly branched inflorescence with stalked individual flowers

*Pappus* — The modified calyx of a disk or rayflower in the sunflower family

*Petaloid* — Petallike

*Ray flower* — Marginal petallike flowers in the inflorescence of the sunflower family

*Rhizome* — A horizontal underground stem

*Rosette* — A cluster of spreading or radiating basal leaves

*Runner* — A horizontal aboveground stem; a stolon

*Sepal* — One of the outer set of floral leaves or calyx, usually green

*Sheath* — The basal portion of the leaf which clasps the stem

*Spike* — An unbranched inflorescence in which the flowers are not stalked

*Spikelet* — The unit of the inflorescence in grasses

*Stamen* — The male portion of the flower

*Stipule* — One of a pair of appendages at the base of the leaf

*Stolon* — A horizontal, underground stem

*Succulent* — Soft and fleshy in texture

*Tuber* — A swollen underground stem

*Volva* — The membraneous envelope enclosing an immature mushroom

*Whorl* — Three or more leaves or flowers arising from the same node

# GENERAL INFORMATION

# GENERAL INFORMATION

Economic losses from weeds are of continuing concern to Arizona farmers and ranchers. Weeds decrease farm income by robbing the soil of precious moisture that would otherwise be available for crop production, by utilizing soil nutrients needed by cultivated plants, by lowering the quality of farm products because of weed impurities, and by increasing the cost of labor, equipment, and irrigation. Many weeds also harbor some of the worst crop insect pests, and are alternate hosts to organisms causing crop diseases. Some weeds are parasites on useful plants.

On rangelands, weeds may seriously decrease grazing capacity for livestock by competing with the good forage plants for moisture and nutrients. Other undesirable range plants may be poisonous, causing reduced weight gains, lowered animal production, or even death. Some range weeds are troublesome to domestic livestock because of their thorns, barbs, stiff hairs, or sharply pointed seeds. Such may cause mechanical injury to the eyes, mouth parts and intestines, and to the skin and the hide. They may lower the value of wool and mohair, or cause wounds that invite attack from screwworms.

Weeds in the yard dampen the home owner's interest in improving and maintaining gardens that beautify the home, and make the community attractive and livable. Weeds also raise the homeowner's costs for water and fertilizer.

## KINDS OF WEEDS

A weed is any plant that grows where it isn't wanted, or is unwanted because of certain undesirable characteristics. Wild roses growing in a pasture are weeds. Bermudagrass is excellent as a summer lawn, but a pernicious weed in flower beds, crop lands, or ditchbanks. Even in lawns, bermudagrass may be undesirable and thus a weed to people afflicted with hay fever. A weed may be almost a personal matter.

Plants that come into crops, irrigated lands, lawns, and gardens; that become established on disturbed soil along roadsides, ditchbanks, neglected fields, and waste places; that invade and increase on rangelands, replacing the cover of perennial grasses or sometimes causing livestock poisoning, are weeds. Such plants include fungi, ferns, grasses, broadleaved herbs, shrubs, and even trees.

CROP, GARDEN AND LAWN WEEDS — Many of the most serious of these weeds in Arizona have been introduced into the United States from the Old World, as wild oat *(Avena fatua)*, nutsedge *(Cyperus rotundus, C. esculentus)*, johnsongrass *(Sorghum halepense)*, junglerice *(Echinochloa colonum)*, hoary cress *(Cardaria draba)*, knotweed *(Polygonum argyrocoleon, P. aviculare)*, field bindweed *(Convolvulus arvensis)*, curly dock *(Rumex crispus)*, and Russian knapweed *(Centaurea repens)*.

Some native plants have also become established as serious pests, as Palmer amaranth *(Amaranthus palmeri)*, silverleaf nightshade *(Solanum elaeagnifolium)*, Texas blueweed *(Helianthus ciliaris)*, alkali sida *(Sida hederacea)*, Wright groundcherry *(Physalis wrightii)*, and red sprangletop *(Leptochloa filiformis)*.

WEEDS OF ROADSIDES, DITCHBANKS, AND WASTE PLACES — May be very aggressive and troublesome. Two of the worst offenders, Russian thistle *(Salsola kali* var. *tenuifolia)*, and puncturevine *(Tribulus terrestris)*, are both Old World introductions.

The common sunflower *(Helianthus annuus)*, crown beard *(Verbesina encelioides* var. *exauriculata)*, alkali heliotrope *(Heliotropum curassavicum)*, slimleaf bursage *(Franseria confertiflora)*, spiny aster *(Aster spinosus)*, and horseweed *(Erigeron canadensis)*, are either native weeds, or were introduced from other parts of the United States.

RANGE WEEDS — Are mostly native plants of low forage value. Opportunity for their increase comes about through overgrazing, drought, and other disturbing influences particularly detrimental to the good forage plants. The junipers *(Juniperus* spp.), mesquite *(Prosopis juliflora)*, burroweed *(Haplopappus tenuisectus)*, broom snakeweed *(Gutierrezia sarothrae)*, and cholla *(Opuntia* spp.), are examples of such native invaders.

POISONOUS RANGE WEEDS — Some plants normally produce some specific substance which, when eaten in sufficient quantity, causes illness or death in livestock; other plants that ordinarily provide good forage may under certain periods of physiological change in the growth pattern produce poisonous substances.

Johnsongrass *(Sorghum halepense)*, patota *(Monolepis nuttalliana)*, Palmer amaranth *(Amaranthus palmeri)*, and Russian thistle *(Salsola kali* var. *tenuifolia)* may, under favorable growth conditions, accumulate high concentrations of nitrate and become poisonous even in small quantities. Johnsongrass may also produce hydrocyanic acid under a variety of environmental conditions, as when growth is stunted, or growth is interrupted and then resumed, as a result of freezing or drought. Other range plants producing hydrocyanic acid are arrowgrass *(Triglochin* spp.), whitethorn, *(Acacia constricta)*, and mountainmahogany *(Cercocarpus* spp.).

The soil environment becomes important when it contains high concentrations of the element selenium. Certain plants, such as species of loco *(Astragalus* spp.), aster *(Aster,* spp. *Machaeranthera* spp.), and saltbush *(Atriplex* spp.), absorb the selenium and may accumulate toxic amounts, causing selenium poisoning in livestock. Such plants have the ability to change insoluble selenium to a soluble form so that other plants growing nearby may also cause poisoning by absorbing the selenium.

3

In most poisonous plants, however, the presence of the toxic principle is not influenced by physiological or environmental variations. There are numerous poisonous compounds which may be present normally in certain plants. The effect of these substances is usually relative to the animal's body weight and the amount of the compound eaten, which often is cumulative. Some of these important stock-poisoning plants in Arizona are loco *(Astragalus* spp.*)*, larkspur *(Delphinium* spp.*)*, western whorled milkweed *(Asclepias subverticillata)*, snakeweed *(Gutierrezia* spp.*)*, cocklebur *(Xanthium* spp.*)*, burroweed *(Haplopappus tenuisectus)*, jimmyweed *(H. plurifolius)*, pinque *(Hymenoxys richardsonii* var. *floribunda)*, and threadleaf groundsel *(Senecio longilobus)*.

Some plants are more poisonous when young, as larkspur *(Delphinium* spp.*)* or cocklebur *(Xanthium* spp.*)*, which is poisonous in the seedling stage. Others, like the lupines *(Lupinus* spp.*)*, are more dangerous in seed. All parts of the plant either green or dry, as in hay, are poisonous in western whorled milkweed *(Asclepias subverticillata)*, loco *(Astragalus* spp.*)*, western sneezeweed *(Helenium hoopesii)*, or western bracken *(Pteridium aquilinum* var. *pubescens)*. Some weeds are about equally poisonous to all kinds of livestock, while others affect only certain kinds of animals. Thus pinque *(Hymenoxys richardsonii)*, western whorled milkweed *(Asclepias subverticillata)*, or lupine *(Lupinus* spp.*)* are harmful chiefly to sheep, and larkspur *(Delphinium* spp.*)*, snakeweed *(Gutierrezia* spp.*)*, or whitethorn *(Acacia constricta)* mainly to cattle.

THE "GOOD" WEEDS — Not all weeds are "bad," but when plants grow in weedy places they suffer from the stigma of being classed as weeds. Many of these are native plants that come in great abundance to the moist, disturbed areas along roadsides and waste places, or that grow in sterile caliche soil, their native habitat, in adjacent mesas, or vacant lots. These weeds are usually not aggressive, nor poisonous to livestock, nor do they invade croplands and thus compete in cultivated areas. The highways throughout Arizona from spring through late summer are lined with colorful weeds that often help to beautify our state. Many unimportant weeds are described here because they are abundant and people often want to know their names. Some of these are: globemallow *(Sphaeralcea* spp.*)*, trailing four-o-clock *(Allionia incarnata)*, woolly tidestromia *(Tidestromia lanuginosa)*, Gordon bladderpod *(Lesquerella gordoni)*, western clammyweed *(Polanisia trachysperma)*, Rocky Mountain beeplant *(Cleome serrulata)*, hairy bowlesia *(Bowlesia incana)*, skeleton weed *(Eriogonum deflexum)*, desert senna *(Cassia covesii)*, and slimpod senna *(C. leptocarpa)*.

## CLASSIFICATION OF WEEDS

Grouped according to their growth habits, there are three principal classes of weeds: annual weeds, biennial weeds, and perennial weeds.

ANNUAL WEEDS — Live one growing season. They flower, produce seed, and then die down entirely. Most annuals, and also biennials, are dependent upon seed alone for reproduction and are usually prolific seeders. There are two general types of annuals: summer annuals and winter annuals.

SUMMER ANNUALS — Are true annuals. They germinate in the spring, mature and produce seeds during the summer, and die in the fall or winter when frost occurs. Some examples are junglegrass *(Echinochloa colonum)*, red sprangletop *(Leptochloa filiformis)*, Russian thistle *(Salsola kali* var. *tenuifolia)*,

Palmer amaranth *(Amaranthus palmeri)*, and puncturevine *(Tribulus terrestris)*. A few summer annuals also produce new plants from prostrate stems rooting at the nodes or joints as large crabgrass *(Digitaria sanguinalis)* and southwestern cupgrass *(Eriochloa gracilis)*.

WINTER ANNUALS — Germinate in late summer or fall and live through the winter as a small rosette of leaves. In the spring they mature rapidly and produce flowers and seeds. Many of these weeds also grow as summer annuals. In central and southern Arizona, such annuals as London rocket *(Sisymbrium irio)*, shepherdspurse *(Capsella bursa-pastoris)*, nettleleaf goosefoot *(Chenopodium murale)*, silversheath knotweed *(Polygonum argyrocoleon)*, spiny sowthistle *(Sonchus asper)*, wild oat *(Avena fatua)*, and many others grow luxuriantly during the short mild winters.

BIENNIAL WEEDS — Live two growing seasons. They germinate in the spring and spend the summer and winter in a rosette form, storing up food in thick roots, usually a taproot. Growth is resumed the next spring, producing a flowering plant which sets seeds, matures, and dies. There are few biennial weeds in Arizona. Common mullein *(Verbascum thapsus)* and golden corydalis *(Corydalis aurea)* are true biennials. Some weeds, commonly annual in colder areas, may become biennial in habit in Arizona, as little mallow *(Malva parviflora)*, camphorweed *(Heterotheca subaxillaris)*, white sweetclover *(Melilotus alba)*, and annual yellow sweetclover *(M. indica)*.

PERENNIAL WEEDS — Live more than two growing seasons. The aboveground parts usually die at the end of the growing season. At the next growing season, new growth arises from the underground structures, which live from one season to the next. In addition to maintaining these old plants, new plants are always produced from seed, and often also from vegetative structures. Based on their methods of producing new plants, most perennial weeds may be classed as simple perennials or creeping perennials. A few possessing unusual structures are classed as specialized perennials.

SIMPLE PERENNIALS — Depend entirely on seed for the production of new plants, except where occasionally pieces of the root crown may be broken off and new plants started. Simple, non-grass perennials usually have a thick taproot topped by a root crown from which new growth arises each growing season. The taproot may be elongated, as in curly dock *(Rumex crispus)*, tuberous sida *(Sida hederacea)*, and dandelion *(Taraxacum vulgare)*; or it may be short and covered by a mass of fibrous side roots as in broadleaf plantain *(Plantago major)* and buckhorn plantain *(P. lanceolata)*. Simple perennial grass weeds have a thick tuft of fibrous roots, as dallisgrass *(Paspalum dilatatum)*, squirreltail *(Sitanion hystrix)*, and foxtail barley *(Hordeum jubatum)*.

CREEPING PERENNIALS — Produce new plants from seed also, but their principal means of reproduction is vegetatively from creeping or horizontal stems and creeping or horizontal roots.

CREEPING OR HORIZONTAL STEMS — Are branches that arise from the lower stem nodes, spreading out horizontally from the plant, and may be underground or aboveground.

Creeping underground stems are called rhizomes or rootstocks. These are true stems, consisting of a succession of nodes or joints, with leaves (usually

**5**

scalelike) on the nodes, and a bud in the axil of each leaf. New plants with leafy shoots and adventitious roots, also the secondary branches, always arise at the nodes and not just at any point as they do in horizontal roots. Creeping stems may be found near the surface or penetrate deeper by gradually sloping downward, never vertically straight down as in roots. Weeds of this type are difficult to eradicate for the rhizomes may store considerable food or extend over large areas. Also, the bud at the node is capable of sending up a new plant even when the rhizome is cut into pieces. Johnsongrass *(Sorghum halepense)* and red sorrel *(Rumex acetosella)* are examples.

Creeping aboveground stems, called runners or stolons, trail along on top of the ground. They are long, slender, and leafy, forming roots at nearly every node; or nearly leafless and taking root only near the tip. These weeds are usually not as hard to eradicate as those with rhizomes. Rocky Mountain iris (*Iris missouriensis*), cinquefoil (*Potentilla anserina*), and white clover (*Trifolium repens*) are examples. A few weeds possess both creeping underground stems (rhizomes) and aboveground stems (stolons), as bermudagrass *(Cynodon dactylon)*.

CREEPING OR HORIZONTAL ROOTS — Are lateral branch roots. These are true roots, arising from a vertical root, usually an elongated taproot, and lack nodes and leaves. Adventitious buds may form at any point along these roots and produce new plants by sending up leafy shoots. Creeping roots can penetrate to a much greater depth than creeping stems. Most of the taproot may extend below cultivating depths, and produce new deeper, horizontal branches when the upper ones are destroyed. Also, the horizontal roots can grow straight downward at any point, thus establishing very deep, extensive root systems, which are virtually impossible to eradicate.

Some of the worst prohibited noxious weeds reproduce by creeping roots, as Russian knapweed (*Centaurea repens*), hoary cress (*Cardaria draba*), field bindweed (*Convolvulus arvensis*), Canada thistle (*Cirsium arvense*), silverleaf nightshade *(Solanum elaeagnifolium),* and blueweed *(Helianthus ciliaris).*

SPECIALIZED PERENNIALS — A few perennial weeds reproduce by means of unusual structures such as tubers, bulbs, and bulblets.

TUBEROUS PERENNIALS — In addition to reproducing by seed and vegetatively by means of creeping stems or creeping roots, a few weeds also produce new plants from small tuberous enlargements formed at the ends of the underground stems or roots. These are storage organs, capable of sending up new plants at any time. They often remain in the ground when the rest of the plant has been grubbed out.

The tubers formed at the ends of underground stems are actually modified rhizomes. Before maturity they are covered by scaly leaves at the nodes, indicating that they are stem structures, as in yellow nutsedge (*Cyperus esculentus*) and purple nutsedge (*C. rotundus*). Tubers formed by creeping roots are clearly root structures, as in hogpotato (*Hoffmanseggia densiflora*). These lack the nodes and scalelike leaves of the stem tubers, even at a very early stage.

BULBOUS PERENNIALS — Usually belong to the lily family, as wild onion. These weeds are common in Arizona, but none are discussed in the text. They seldom reproduce by seed, but rather by the multiplication of bulbs and bulblets (tiny bulbs) under the ground. Aerial bulblets may also be formed, often in the leaf axils.

## PROHIBITED AND RESTRICTED NOXIOUS-WEED SEED

Noxious-weed seeds are so declared by law. They may be divided into two classes: prohibited noxious-weed seed — produced by weeds which are so highly detrimental and difficult to control that the sale of commercial planting seed which contain any of these seeds is prohibited; and restricted noxious-weed seed — from very objectionable weeds, but more easily controlled, thus a limited amount of such seed is permitted. Each of the states has its own seed laws in which the "prohibited noxious-weed seed" and the "restricted noxious-weed seed" are listed by both their common and scientific names. In addition, the maximum amount of restricted noxious-weed seed permitted is stated. The laws also require that seed labels state the quantity per ounce or per pound of any restricted noxious-weed seed present.

PROHIBITED NOXIOUS-WEED SEED IN ARIZONA — The Arizona Seed Law designates the following as "prohibited noxious-weed seed" in Arizona: (The plants with asterisks are not known to date in Arizona, but are in nearby states. By prohibiting their seed in commercial seed, the chance of their entry is greatly reduced.)

Bindweed, field (*Convolvulus arvensis* L.)
Blueweed, Texas or blueweed (*Helianthus ciliaris DC.*)
Camelthorn (*Alhagi camelorum* Fisch.)
Cress, hoary or whitetop (*Lepidium draba* L., *Cardaria draba* Desv., or
 *\*L. repens* Boiss.)
\*Horsenettle (*Solanum carolinense* L.)
Horsenettle, white or silverleaf nightshade (*Solanum elaeagnifolium* Cav.)
Knapweed, Russian (*Centaurea repens* L., or *C. picris* Pall.)
Morningglory (*Ipomoea* spp.)
Nutsedge, purple (*Cyperus rotundus* L.)
Nutsedge, yellow (*Cyperus esculentus* L.)
Perennial sorghum (*Sorghum* spp.) as johnsongrass (*S. halepense* L.),
 sudangrass (*S. sudanese* [Piper] Stapf.), and \*sorghum almum (*S.
 almum* Parodi)
\*Sowthistle, perennial (*Sonchus arvensis* L.)
\*Spurge, leafy (*Euphorbia esula* L.)
Thistle, Canada (*Cirsium arvense* (L.) Scop.)
\*Whitetop, hairy or hoary cress (*Cardaria pubescens* (C. A. Mey.) Roll.
 var. *elongata* Roll., or *Hymenophysa pubescens* C. A. Mey.)

RESTRICTED NOXIOUS-WEED SEED IN ARIZONA — Seeds of the following weeds are declared as "restricted noxious-weed seed" in Arizona: (The numerals after each plant name indicate the maximum number of these particular seeds allowed per pound in commercial planting seed. A total of 500 of all such seed per pound is allowed.)

| | |
|---|---:|
| Dock, curly (*Rumex crispus* L.) | ——300 |
| Dodder (*Cuscuta* spp.) | —— 45 |
| Mallow, alkali (*Sida hederacea* [Dougl.] Torr.) | ——300 |
| Mustard, wild or wild turnip (*Brassica* spp.) | ——300 |
| Oat, wild (*Avena fatua* L.) | ——100 |
| Puncturevine (*Tribulus terrestris* L.) | ——100 |
| Sandbur (*Cenchrus pauciflorus* Benth.) | ——100 |
| Thistle, Russian (*Salsola kali* L. var. *tenuifolia* Tausch.) | ——300 |

## WEED NAMES

All weeds have both a common name and a scientific name. A single common name for a given weed has been recommended by the Weed Society of America in their report of Common and Botanical Names of Weeds. This recommended name is printed in capital letters in this book. Additional common names applied to a given weed are also listed, not as a choice of common names, but to facilitate the weed's identification.

The use of common names has resulted in such confusion that it is necessary to use a plant's scientific name. No laws cover the usage of common names; the same common name is often given to a variety of weeds, particularly closely related species. Also, most weeds have numerous common names. To end this confusion, the name recommended by the Weed Society should be universally adopted, even though it might not always be as popular as another. Now that each particular weed species has been given a precise common name, it should be used.

A plant scientific name is governed by strict international botanical law and is an exact name for one particular species. Sometimes a name must be changed due to increased knowledge. The original name then becomes a synonym of that species, and though outmoded, it still refers to that species and cannot be applied to another. The author's name is an essential part of the scientific name, and is abbreviated at the end of the Latin name.

For the sake of uniformity, the same scientific name as given in *Arizona Flora* is used unless another name is now the accepted one. (The older name is then placed in parentheses.)

## IDENTIFICATION OF WEEDS

The purpose of this book is to assist the farmer, rancher, homeowner, or agricultural student to recognize common Arizona weeds. Most weeds may be identified by comparison with the drawings, and checking the plant description. To identify a weed not described in the book, use Kearney & Peebles, *Flora of Arizona,* or send the weed to the Herbarium, University of Arizona, Tucson, for free identification. The specimen should be pressed and dried before sending , and should include all parts — roots, stems, leaves, flowers, and fruits.

# WEED SPECIES OF ARIZONA

# FLY MUSHROOM, fly amanita, fly agaric
## MUSHROOM FAMILY — Agaricaceae

MUSHROOM FRUITING BODIES — The familiar mushroom, consisting of a stalk and an umbrellalike cap, is the fruiting body of these fungus plants. These arise at irregular intervals from the buried vegetative parts, usually after prolonged periods of summer rainfall. On the upper part of the stalk is the ring, the remains of an early veil that extended from the cap to the stalk and later ruptured, leaving a remnant on the stalk.

### FLY MUSHROOM — Amanita muscaria L.

DESCRIPTION — The colorful fruiting body of fly mushroom is very conspicuous. The mature cap varies in color from brilliant scarlet to dark red, reddish-orange to orange, or yellow-orange to pale yellow, and in width from 2½ to 10 inches. The upper surface of the cap is covered with many wartlike patches; white at first, yellowish-brown in age. Both the gills and spores (in mass) are white. The whitish stalk, faintly yellow in age, is 4 to 8 inches high, with a bulbous base. The large, deflexed, whitish ring may be lost in age.

DISTRIBUTION — Fly mushroom is found in woods throughout the holarctic. In Arizona it is common in the mountains along the tree margins of meadows or forest openings. It lives in the soil humus, and becomes attached to the outside of rootlets of certain forest trees, forming mycorrhizae. Mycorrhizal relationships may be established with pine, spruce, fir, and aspen. Like greenspored mushroom, the fruiting bodies are most common from August to October, and sometimes form "fairy rings."

POISONOUS PROPERTIES — After the fly mushroom was eaten, the following symptoms have been known to occur: The nervous system is seriously affected. Retching, vomiting, excessive salivation, watery diarrhea, perspiration, and hallucinations usually appear within ½ to 3 hours. Sometimes death follows through respiratory failure, preceded by delirium and convulsions. Cattle may be poisoned by fly mushroom, but no losses, either of human or livestock, are known in Arizona.

The most deadly poisonous mushrooms known belong to the genus *Amanita*. Most species are distinguished by the remnants of an outer enveloping veil, or volva, found around the base of the stem and in patches on top of the cap. The volva at the stem base may form a baglike cup (death cup); or a narrow rim above a bulbous base; or a series of scaly circles on the stem above a bulbous base, as in fly mushroom. Characteristically, the ring on the upper part of the stem hangs like a short skirt.

10

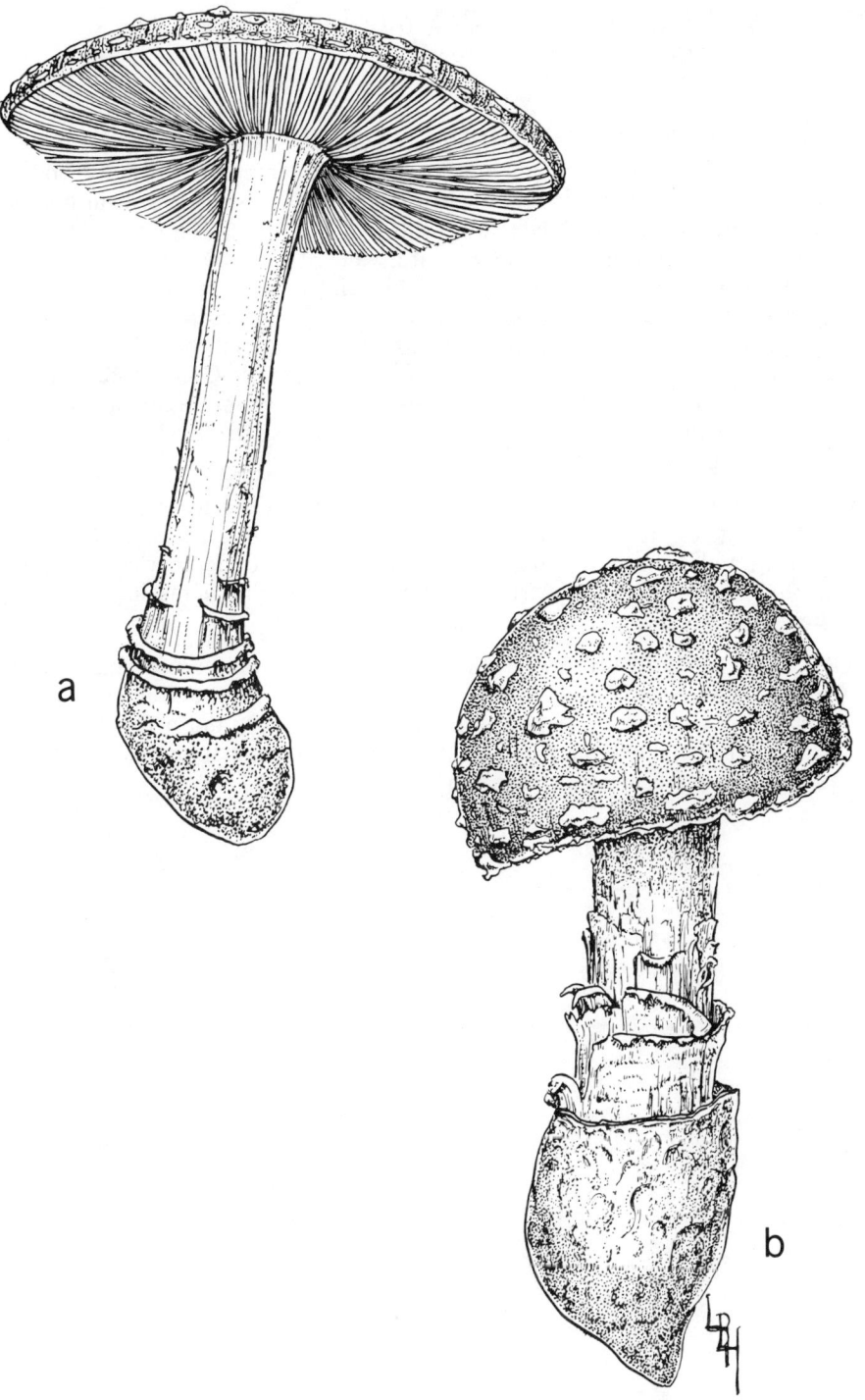

Fig. 1. Fly mushroom *(Amanita muscaria). a.* Cap and stalk of mature fruiting body. Whitish warts cover the cap's upper surface and gills the lower. The pendulous ring is shown on the upper part of the stalk; on the lower are the scaly circles of volva remnants above the bulbous base. *b.* Young fruiting body.

## GREENSPORED MUSHROOM, morgan lepiota,
### greenspored lepiota
MUSHROOM FAMILY — Agraricaceae

### GREENSPORED MUSHROOM — *Lepiota morganii* **Pk.**

DESCRIPTION — Greenspored mushroom has a large thick fruiting body. The white cap is 4 to 10 inches broad when mature, with irregular brownish scales on top. The greenish color of the mature gills and spores is a distinctive characteristic of greenspored mushroom. The gills are white at first, but turn green when the spores begin to fall. The spores are bright green when first mature, but fade to dull green and finally brownish black. The stalk is white, grayish-white, or tinged with brown, 4 to 8 inches high, and tapers upward from a clubshaped base. The large ragged-edged ring is thick, movable, and not deflexed. Greenspored mushroom looks very much like the common cultivated mushroom *(Agaricus campestris),* but the gills in the latter are pinkish and the spores purplish-brown at maturity, rather than green.

DISTRIBUTION — Greenspored mushroom is found mostly in the western and southern states. It is common in organic soil in yards, fields, and pastures in southern and central Arizona; usually from August to October. In some places, successive years of crops of fruiting bodies may occur in increasingly larger rings, often 16 feet or more in diameter, forming "fairy rings."

POISONOUS PROPERTIES — Some people can eat this mushroom with no ill effects; others become violently ill with vomiting, diarrhea, and intense abdominal pain within an hour or less after eating a small piece, and a few have died. In Arizona there are no known cases of poisoning, either human or livestock, definitely attributable to greenspored mushroom.

Fig. 2. Greenspored mushroom *(Lepiota morganii).* *a.* Mature fruiting body showing umbrellalike cap, the top scaly, the underside with gills; and the stalk with its firm torn-edged ring near the cap. *b.* Young fruiting body.

## WESTERN BRACKEN, bracken, bracken fern, brake fern, eagle fern

FERN FAMILY — Polypodiaceae

### WESTERN BRACKEN — *Pteridium aquilinum* (L.) Kuhn var. *pubescens* Underw.

DESCRIPTION — A perennial fern which reproduces by spores and widely creeping, branching underground stems, sometimes forming colonies. The large compound leaves (fronds) are 1 to 4 feet high, and ½ to 1½ feet long. The leaf stalk, usually mistaken for the stem, actually is attached to the rhizome under the ground.

The triangular deciduous leaves turn brown and die at fall frost, and the new ones arise each spring from the rhizomes. The leaf is divided into numerous segments (leaflets), each of which may be again divided or redivided, with the lowest segments three times compound. The clusters of spore cases densely line the inrolled edges of the underside of the leaves; the spores mature usually in July and August.

DISTRIBUTION — Western bracken, a native and our only weedy fern, grows in the mountains in Arizona. It is very common in shade or partial sun in open yellow pine forests, less common in aspen and spruce; it is also found along the tree edges of mountain meadows, and on steep slopes at higher elevations. It comes in quickly on forestlands which have been burned or logged. In Apache, Navajo, and Coconino counties, southward to Cochise and Pima counties at 5,000 to 8,500 feet elevation.

POISONOUS PROPERTIES — In early spring the young unrolled leaves and tender leaf stalks may be cooked as a vegetable. When mature and tough, they are poisonous to horses and cattle. Sheep have been poisoned experimentally, but natural poisoning is unknown in the United States, and goats are immune. The rhizomes are five times more toxic than the leaves, but are seldom eaten.

The poison is cumulative over a period of about 1 month for horses, and 1 to 4 months for cattle before symptoms appear. Horses usually are poisoned by eating large amounts of hay containing over 20% of dry bracken. Cattle are poisoned by an amount of green or dried leaves about equal to the animal's weight. Livestock, however, seldom eat western bracken on the range when other forage is available.

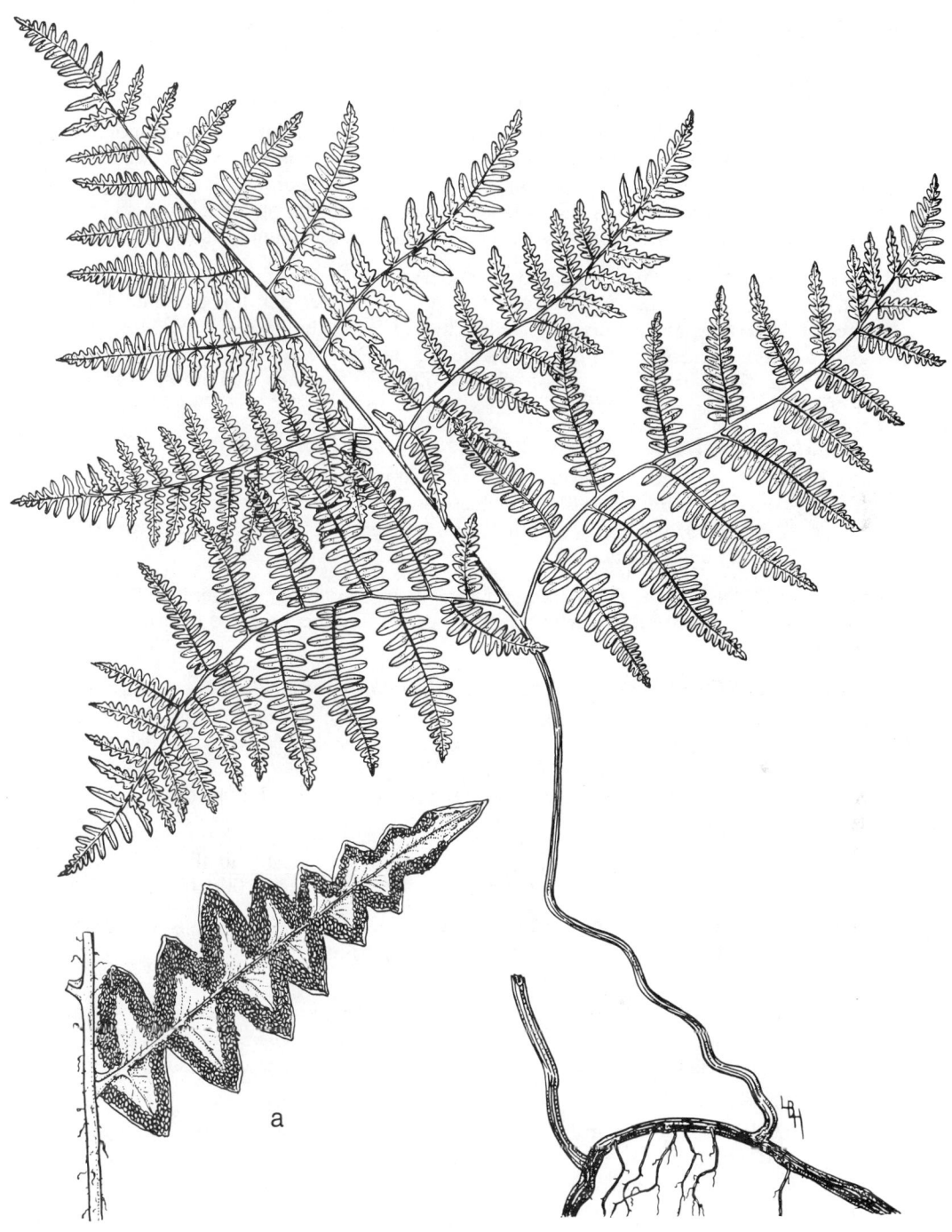

Fig. 3. Western bracken *(Pteridium aquilinum* var. *pubescens).* The
frond with its slender stalk attached to the rhizome. *a.* The abundant
clusters of spore cases on the inrolled margins of the underside of the leaf.

# JUNIPER

CYPRESS FAMILY — Cupressaceae

### JUNIPER, redcedar — *Juniperus* spp.

DESCRIPTION — Aromatic trees or shrubs with short overlapping sharp-pointed leaves, about 1/16 inch long with green berrylike cones; reproducing by seed and often by stump sprouts. The "berries," ¼ to ½ inch in diameter, are relished by birds and other wild animals. Livestock may eat the foliage, but usually do so only when other feed is scarce. Large amounts may cause abortion in livestock.

DISTRIBUTION — In northern and central Arizona below the ponderosa pines, junipers often form pure forests or mixed pinyon-juniper-oak associations. They are a serious range and watershed problem in large grassland areas on the Mogollon Rim north of Prescott, and on the Apache Indian Reservation. Junipers are long-lived trees; the average age is about 500 to 600 years, and the maximum is about 1,600 years.

### UTAH JUNIPER — *Juniperus osteosperma* (Torr.) Little

DESCRIPTION — Utah juniper usually is a small round tree 10 to 15 feet high with a single trunk, or if branched, the branches arise several feet above the ground. The "berries" contain 1 or 2 seeds, and may be produced on all of the trees. Its wood is used in making fence posts.

DISTRIBUTION — The most abundant juniper in the state, Utah juniper is wide-spread over northern and central Arizona, but does not occur in the southern part; 3,000 to 7,500 feet elevation.

### ONESEED JUNIPER — *Juniperus monosperma* (Engelm.) Sarg.

DESCRIPTION — Oneseed juniper usually is shrubby, 10 to 25 feet high with several curved limbs arising near the ground level, and only rarely a tree with a single trunk. Like Utah juniper, the bark is stringy and the "berries" contain 1 or 2 seeds, but unlike it, the "berries" and the pollen occur on different trees.

DISTRIBUTION — Common in the desert grassland and pinyon-juniper ranges in southeastern and northcentral Arizona, but does not occur in the western or upper northern portions; often growing with Utah juniper; 3,000 to 7,000 feet elevation.

### ALLIGATOR JUNIPER, checker bark juniper — *Juniperus deppeana* Steud.

DESCRIPTION — Alligator juniper, the largest in Arizona, is a tree 20 to 40 or sometimes 65 feet high, with a trunk 1 to 3 or rarely 7 feet in diameter. It often stump sprouts, and the thick, dark gray bark becomes deeply checkered. The "berries" contain 3 or 4 seeds.

DISTRIBUTION — Well-distributed in central and southeastern Arizona, alligator juniper extends from southwestern Coconino and northwestern Yavapai counties to Greenlee County, and west to the Baboquivari Mountains in Pima County; 4,200 to 8,000 feet elevation.

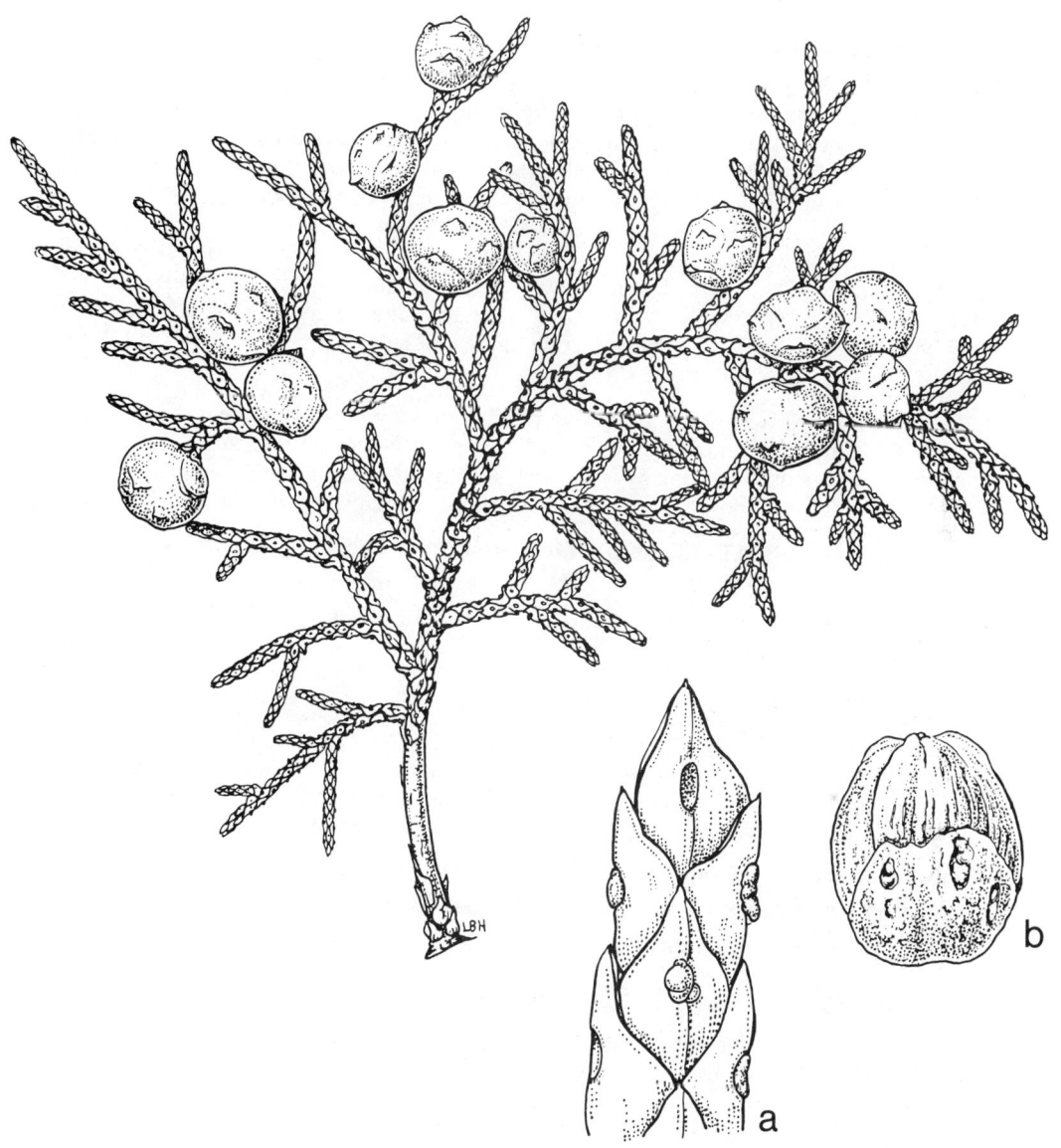

Fig. 4. Alligator juniper *(Juniperus deppeana)*. Branch with
berrylike female cones. *a.* Mature appressed-scalelike leaves
in alternate pairs, with glandular pits on the back. *b.* Seed.

17

# CATTAILS

CATTAIL FAMILY — Typhaceae

### CATTAILS — *Typha* spp.

DESCRIPTION — The two most common species of cattail in Arizona are stout pithy perennials, 4 to 7 feet high, reproducing by seed which germinates readily on wet mud; also spreading rapidly by creeping, submerged rhizomes. The alternate grasslike leaves are several feet long, with sheathing bases. The "cattail" is the flowering spike, composed of two sections of minute flowers.

The upper yellowish section bears only male flowers, each consisting of 2 to 5 stamens; the flowers drop away as soon as the pollen is shed, leaving the upper end of the spike naked. The lower brown section bears masses of densely packed female flowers, each with 40 to 60 delicate hairs, and persists for several months. The seedcase is a minute achene, brown and spindleshaped, about 1/16 inch long, with the fine hairs still intact when shed.

DISTRIBUTION — Cattails are native plants, growing in marshy areas along rivers and floodlands, but may become a nuisance by clogging irrigation ditches and permanent ponds; flowering May to July. With their efficient vegetative propagation they form dense stands, and are hard to eradicate once established.

### COMMON CATTAIL, soft flag, broadleaf cattail — *Typha latifolia* L.

DESCRIPTION — The leaves flat, ⅜ to ⅞ inch broad, and usually without a space between male (upper) and female (lower) sections of spike. The female section is fat-cigarshaped, dark reddish brown, whitish in age, 4 to 7 inches long, ¾ to 1¼ inches thick, and often somewhat thicker at top.

DISTRIBUTION — Common cattail usually grows in fresh water in slightly acid soils, sometimes in slightly polluted water, or in seepage areas, even on slopes. Mostly in north central and eastern Arizona, in Apache, Navajo, Coconino, Gila, Maricopa, and Cochise counties; 3,500 to 7,500 feet elevation.

The leaves are woven into chair seats in central New York.

POISONOUS PROPERTIES — Common cattail has been suspected of fatally poisoning horses, but no cases have been reported in Arizona.

### SOUTHERN CATTAIL — *Typha domingensis* Pers.

DESCRIPTION — The leaves are somewhat convex on the back, and ¼ to ½ inch broad. A space from ¼ to 1¾ inches long is always present between the male and female sections of the spike. The female section is a lighter color than the common cattail, light brown becoming buffy or grayish, longer and narrower, 6 to 15 inches long, ⅝ to ⅞ inch thick and the same thickness throughout.

DISTRIBUTION — Southern cattail grows mostly in lowlands in brackish water or wet soils with appreciable salt content; in north central and southern Arizona, in Apache, Coconino, Gila, Maricopa, Pinal, Cochise, Pima, and Yuma counties; 137 to 6,500 feet elevation.

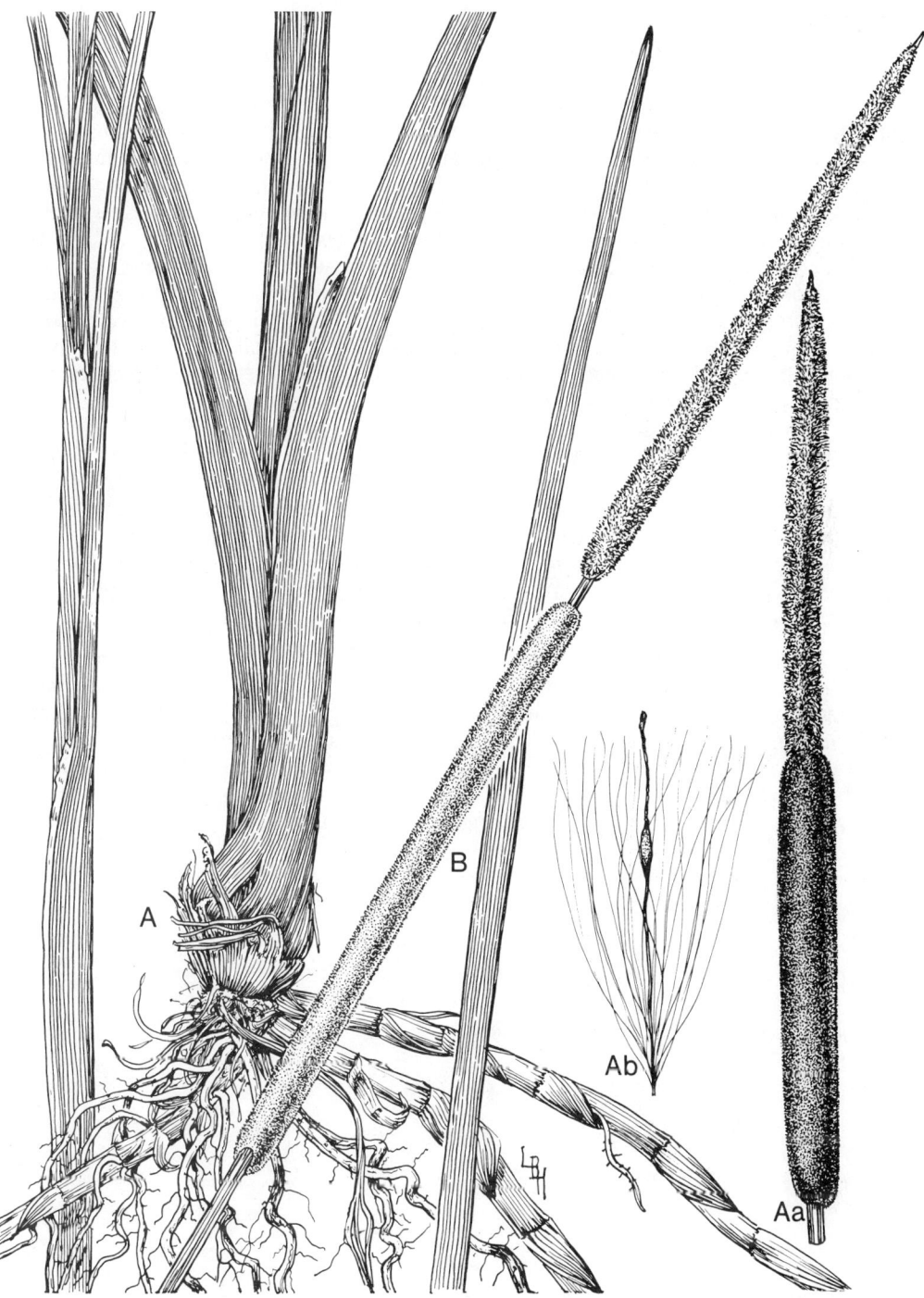

Fig. 5. *A.* Common cattail *(Typha latifolia).* Plant showing rhizomes.
*Aa.* Flowering spike; the upper section of male flowers, the lower, females.
*Ab.* Achene. *B.* Southern cattail *(T. domingensis).* Flowering spike and leaf.

19

# LEAFY PONDWEED

PONDWEED FAMILY — Potamogetonaceae

### LEAFY PONDWEED — *Potamogeton foliosus* Raf.

DESCRIPTION — Perennial aquatics from slender rhizomes; reproducing by seed, rooting at the nodes of the rhizomes, and by winterbuds. The threadlike stems, little branched below, but bushy branched above, are up to 3½ feet high, varying according to the depth of the water. The alternate narrowly linear leaves are all submerged, deep green to bronze, the principal ones to 1/10 inch broad, and to 4 inches long, have 3 to 5 veins, and the stipules at the leaf bases are ¼ to ¾ inch long. In the var. *macellus,* also found in Arizona, the leaves are bright green, the principal ones to 1/18 inch broad, and ¼ inch long, have 1 to 3 veins, and the stipules are ⅛ to ½ inch long. The winterbuds are stalkless in the leaf axils or at the end of very short branches, but in the var. *macellus* the branches are longer.

The tiny flowers have 4 brownish petallike parts, only 1/25 inch long, are stalkless, and arranged in pairs of 2 to 12 flowers, in short headlike flower clusters (spikes). The few flower spikes, ⅛ to ¼ inch long, in the upper forks of the stems are on stalks up to 1 inch long, and are slightly thickened upward. The flower clusters are above the water surface while in flower for wind pollination, but submerged by the time the fruits are mature. The nutlets are greenish brown, unequally circular, about 1/10 inch in diameter, with a prominent knife-edged, often scalloped keel on its back. The beak is erect with a broad base.

DISTRIBUTION — Leafy pondweed grows in rather shallow, still, slowly running, or swift streams, lakes, or ponds; in fresh and alkali, usually hard, or sometimes brackish water. The entire plant is submerged except the flowers. It has become a nuisance in the irrigation ditches and reservoirs in the commercial cropland areas in Arizona. Often the plants are so dense as to actually retard the water flow. Found in Coconino, Mohave, Maricopa, Gila, Pima, and Santa Cruz counties; 1,000 to 8,500 feet elevation; flowering July to October.

a

b

Fig. 6. Leafy pondweed *(Potamogeton foliosus).* Plant with threadlike stems and leaves; also the tiny flowers in very small flowering spikes. *a.* Flower showing four petallike parts. *b.* Nutlet with scalloped keel and erect beak.

21

## WILD OAT, oatgrass, wheat oats, flaxgrass
### GRASS FAMILY — Gramineae

### WILD OAT — *Avena fatua* L.

DESCRIPTION — Wild oat is a restricted noxious-weed in Arizona. It is a stout annual, 1 to 4 feet high, with a large root system, and reproduces by seed. The leaf blades, 3 to 8 inches long, and about ¼ to ½ inch broad, are thin and rough. The flowering part is large and spreading. Each spikelet is about 1 inch long without the bristle, and has 2 to 4 flowering bracts. These bracts have reddish-brown hairs at the base and, arising from the center back of each is a dark, stout, 1¼ to 1⅔ inch long bristle. The bristle is bent sharply near the center, and the lower part is twisted slightly. The white or yellowish to gray, brown, or black grain is silky, hairy, and about ⅓ inch long.

DISTRIBUTION — Wild oat is one of the most troublesome winter crop pests in the state. It has been estimated that most small field grains in the Salt River Valley vary in content from 1 percent to 75 percent of wild oat. It is a common weed in disturbed soil and waste places throughout most of the state; up to 8,250 feet.

### CULTIVATED OAT — *Avena sativa* L.

DESCRIPTION — Cultivated oat and its varieties closely resemble wild oat, but the bases of the flowering bracts are not hairy, and if a bristle is present, it is shorter and straight, not bent in the center.

22

Fig. 7. Wild oat *(Avena fatua)* plant and flowering branch.
*a.* Spikelet. *b.* Floret. *c.* Grain.

23

# SIXWEEKS NEEDLE GRAMA, needle grama
GRASS FAMILY — Gramineae

### SIXWEEKS NEEDLE GRAMA — *Bouteloua aristidoides* (H. B. K.) Griseb.

DESCRIPTION — A low, delicate, short lived summer annual, reproducing by seeds. The slender stems are 3 to 12 or more inches high, from a weak shallow root. The leaves are thin, few, 3 to 6 inches long, and about 1/12 inch broad.

There are 4 to 15 (rarely 20) flower spikes, very narrow below, with the base sharp pointed and the top spreading, ⅜ to ¾ inch long. They are very loosely attached along one side of the stem, and drop off easily. There are 3 short bristles, about ⅕ inch long, from the tip of 1 or more of the flowering bracts in each spike. The narrow flattened grain is brownish, and about ⅛ inch long.

DISTRIBUTION — Sixweeks needle grama is a native grass which has become weedy. It is found in the same habitat types throughout Arizona, except in the northeastern part. On denuded areas after the start of the summer rains it sometimes occurs in dense stands; 100 to 5,000 feet elevation, but mostly lower than 5,500 feet; flowering from June to October.

Fig. 8. Sixweeks needle grama *(Bouteloua aristidoides).* Plant with very narrow flower spikes. *a.* Two spikelets. *b.* Lowest spikelet. *c.* Grain.

25

# SIXWEEKS GRAMA

### GRASS FAMILY — Gramineae

## SIXWEEKS GRAMA — *Bouteloua barbata* Lag.

DESCRIPTION — An annual bunchgrass, branching from the base, from a shallow weak root, reproducing by seeds. The slender stems may be erect or prostrate, sometimes forming mats with the ends ascending, mostly less than 12 inches high, but may be much higher. The leaves are scarce, short, ⅜ to 1½ inches long, and narrow, 1/16 inch or less broad.

There are 4 to 7 comblike flower spikes on each stem. These are persistent, and do not drop off easily. They are ⅜ to ¾ inch long, with 25 to 40 spikelets arising from just one side of the spike stem; thus giving the spikes the characteristic comb-like appearance. Some of the flowering bracts are tipped by slender bristles, ⅛ to 1/12 inch long.

DISTRIBUTION — Sixweeks grama is a native grass of dry, disturbed, rocky, sandy, or caliche soil. It is a common weed throughout the state in waste places, roadsides, city streets, and in all kinds of summer or fall cultivated fields and orchards in the agricultural areas of southern Arizona. It also is abundant on mesas, hillsides, washes, and barren eroded places, on overgrazed or deteriorating ranges in the desert, desert grassland, or chaparral woodlands; 100 to 6,000 feet elevation, but mostly at the lower elevations; flowering principally from July to September or October, but throughout the year in good locations. This plant and sixweeks needle grama are alternate hosts for the beet leafhopper.

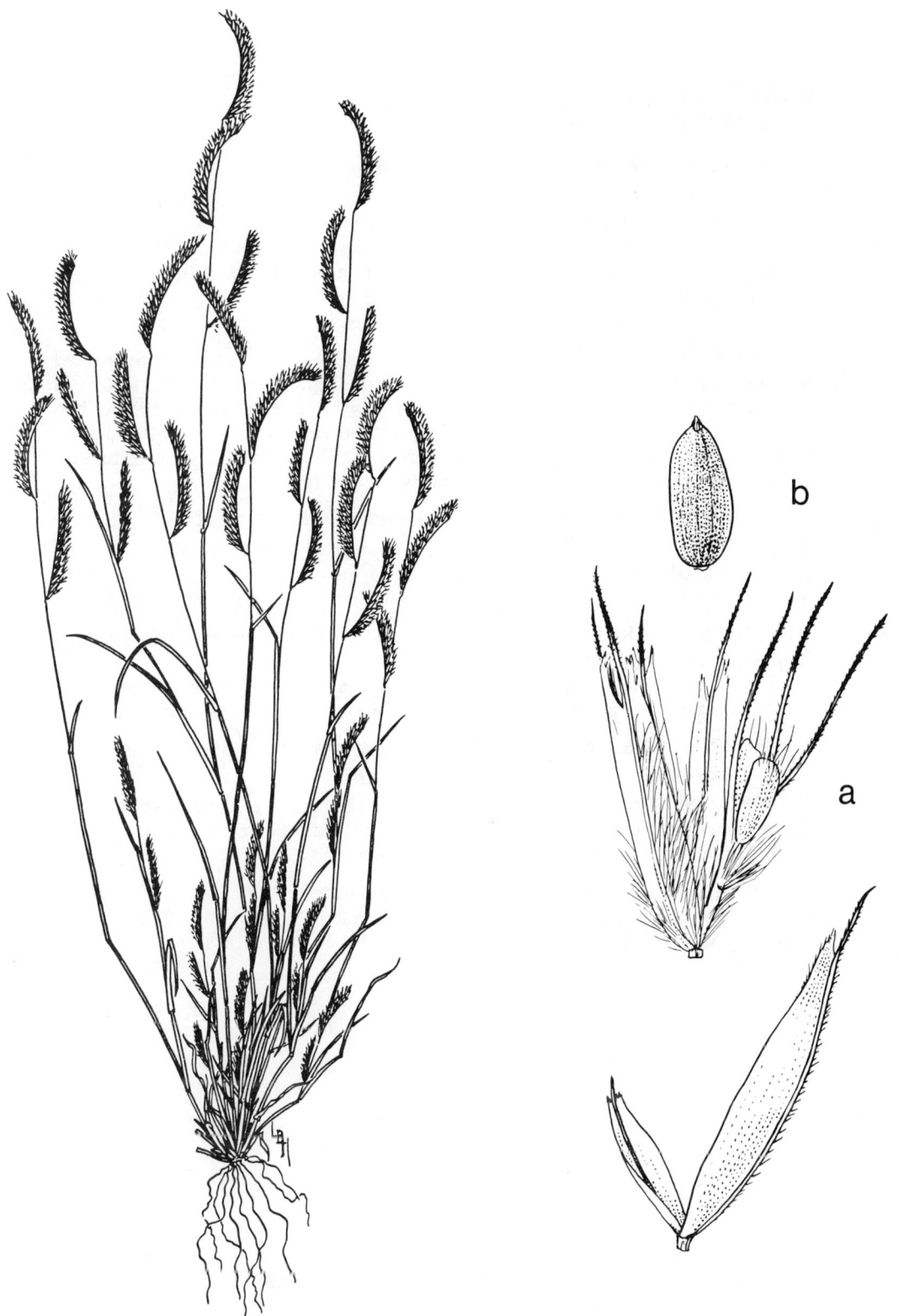

Fig. 9.  Sixweeks grama *(Bouteloua barbata)*.  Plant with comblike spikes.
*a.* Spikelet showing a fertile and sterile flower, with the separated glumes below.  *b.* Grain.

27

# RESCUE BROME, rescue grass
GRASS FAMILY — Gramineae

### RESCUE BROME — *Bromus catharticus* Vahl.

DESCRIPTION — Rescue brome is an annual, a winter annual, or biennial, ½ to 3 feet high, which reproduces only by seeds. It has thin flattish leaves, ⅛ to ⅓ inch broad. The flowering part is branched, and up to 8 inches long. The spikelets are large, ¾ to 1¼ inch long, strongly flattened and composed of 6 to 12 sharply folded, overlapping flowering bracts, which may or may not end in a stiff bristle. The grain is about ⅜ inch long.

DISTRIBUTION — A native of South America, rescue brome is principally a weed in lawns, gardens, roadsides, ditchbanks, and small grain winter crops. Found in all counties, from 140 feet at Yuma to over 7,000 feet elevation on the Kaibab plateau, it is locally abundant around Tucson and rapidly increasing throughout southern Arizona; flowering from late February to September (higher altitudes). Like the bromes below, it disappears during hot weather.

### RED BROME, foxtail brome, foxtail chess — *Bromus rubens* L.

DESCRIPTION — A spring annual, ½ to 1⅓ feet high with the flowering part erect, 1½ to 3½ inches long and crowded. The spikelets, about 1 inch long, have 4 to 11 flowering bracts, each ending in a reddish brown bristle about ¾ inch long.

DISTRIBUTION — Red brome is troublesome, principally on overgrazed rangelands in central Arizona where it has spread rapidly, uncommon southward and northward, 1,300 to 5,500 feet elevation; flowering March to June. The mature bristles are mechanically injurious to the eyes, mouths, and wool of animals.

### DOWNY CHESS, cheatgrass brome, downy brome, slender chess — *Bromus tectorum* L.

DESCRIPTION — Weak, tufted annuals with very hairy leaves. The flowering part, 2 to 6 inches long, is soft, drooping, often purplish, with the flower groups nodding on threadlike stalks, and the bristles ⅜ to ⅝ inch long.

DISTRIBUTION — Downy chess is a European introduction that is primarily a weed on run-down ranges, but in northern Arizona, 4,800 to 8,500 feet elevation; flowering May to July.

Fig. 10. Rescue brome *(Bromus catharticus)*. Plant and flowering branch. *a.* Spikelet. *b.* Grain.

# FIELD SANDBUR

GRASS FAMILY — Gramineae

## FIELD SANDBUR — *Cenchrus pauciflorus* Benth.

DESCRIPTION — Field sandbur is a restricted noxious-weed in Arizona. It is an erect or spreading annual, or sometimes a short-lived perennial, which reproduces by seeds, and by prostrate stems rooting at the lower nodes. It sometimes forms mats, then ascends, 4 inches to 3 feet long, with a shallow root system. The leaf blades are usually flat, twisted or folded, and 2 to 5 inches long. The flowering spikes are 1½ to 4 inches long, often partially enclosed by the upper leaf sheath, and are composed mostly of 3 to 15 burs, loosely arranged, but may have 20 to 30 burs, and be very tightly congested.

The spiny, hairy yellowish burs are about ½ inch long, and mostly longer than broad. The numerous flattened spreading rigid spines, ⅛ to ¼ inch long, often have a few curved bristles at the base. Each bur usually contains 2 seeds. There may be as many as 1,000 seeds produced by a single plant.

DISTRIBUTION — Field sandbur is a native American weed, and is very troublesome when the burs mature. These injure clothing as well as skin. This weed is found mostly in dry sandy soil in cultivated fields, roadsides, lawns, washes, and waste places throughout the state, 100 to 6,000 feet elevation; flowering May to October, or until fall frosts.

## SOUTHERN SANDBUR — *Cenchrus echinatus* L.

DESCRIPTION — Southern sandbur closely resembles field sandbur, but the burs are reddish, mostly broader than long, and are broadest at the base. (Those of field sandbur are broadest near the center.) These have a definite ring of many short, straight bristles at the base. Each bur usually contains 4 seeds.

DISTRIBUTION — Southern sandbur is an introduction from tropical America. It is not as widespread throughout the state as field sandbur, but it is a very troublesome weed in the late summer crops, orchards, and alfalfa fields in the Yuma, Salt River, and Santa Cruz valleys, 100 to 4,000 feet elevation; flowering May to October, or until fall frosts.

30

Fig. 11. Southern sandbur *(Cenchrus echinatus)*. Plant and enlarged branch with inflorescence. *a.* Bur with four spikelets. *b.* Grain, upper and lower view. *c.* Bur of field sandbur *(C. pauciflorus)*.

31

# FEATHER FINGERGRASS
GRASS FAMILY — Gramineae

## FEATHER FINGERGRASS — *Chloris virgata* Swartz.

DESCRIPTION — Feather fingergrass is a weak, slender, erect or spreading annual, ½ to over 3 feet high, with a shallow root system, and reproduces by seeds. The leaf blades are weak, ¾ to 3 inches long, with the upper leaf bases enlarged, and enclosing the flowering part until it opens. The flowering part consists of 2 to 10 tawny, narrowly featherlike soft flowering spikes, ¾ to 3½ inches long, arising together at the stem tips. Each spikelet has a tuft of long soft hairs at the top, and very slender bristles ¼ to ⅜ inch long. The little spikelets are crowded, and arranged in 2 rows along one side of the spike stem.

DISTRIBUTION — This South American weed is very common on bare disturbed soil throughout Arizona in waste places, roadsides, desert washes, and swales, up to 5,500 feet elevation; flowering April to November. It has spread rapidly to the irrigated fields, and is now abundant in cultivated crops in many areas.

## TUMBLE WINDMILLGRASS — *Chloris verticillata* Nutt.

DESCRIPTION — Closely related to feather fingergrass, but a low perennial. The flowering spikes are longer (3 to 6 inches long), stiff, and outward spreading. They arise from 3 or more levels near the top of the stem, and are not feathery. This Texan weed has become established in the Yuma mesa in the citrus orchards.

Fig. 12.　Feather fingergrass *(Chloris virgata)*. Plant and flowering stem.
*a.* Spikelet with glumes in detail below.

33

# BERMUDAGRASS, devilgrass

GRASS FAMILY — Gramineae

## BERMUDAGRASS — *Cynodon dactylon* (L.) Pers.

DESCRIPTION — A wiry, spreading perennial which reproduces by seeds, but mostly by means of long runners on top of the ground. Also by a vast system of hard, sharp-pointed rhizomes beneath the ground which may be shallow or very deep. The leaves on the erect stems are 1 to 4 inches long, while those on the runners and rhizomes are very short, scaly, and not leaflike.

Roots are formed at the joints, and frequent erect, flowering stalks are produced, about 4 to 18 inches high. These have 3 to 5 very narrow fingerlike flowering spikes at the tip. The tiny stalkless spikelets slightly overlap one another, and are arranged in two rows on just one side of the spike axis. The tiny, oval grain is orange red, reddish brown, or straw colored.

DISTRIBUTION — Bermudagrass is one of the most pernicious weeds throughout the state, and is very alkali-resistant. It grows almost any place in cities, waste places, and roadsides — wherever there is moisture. It even occurs in sandy washes of out-of-the-way canyons; usually below 6,000 feet elevation; flowering from May until November.

This troublesome weed is very hard to eradicate when it becomes established in flower beds, noncultivated crops, and fields, and its pollen is one of the most serious sources of hay fever in the state. However, it is the most common summer lawn grass in southern Arizona. It cannot stand freezing temperatures, shade, or frequent cultivation, but can tolerate indefinite periods of drought. Around Yuma, large fields are grown as seed crops.

Fig. 13.  Bermuda grass *(Cynodon dactylon)*. Plant showing rhizomes.
*a.* Enlarged inflorescence.  *b.* Spikelet, with glumes detached below.  *c.* Grain.

## LARGE CRABGRASS, hairy crabgrass

GRASS FAMILY — Gramineae

### LARGE CRABGRASS — *Digitaria sanguinalis* (L.) Scop.

DESCRIPTION — A weak branched summer annual which reproduces by seeds, and by stems spreading at the base and rooting at the lower joints. The flat leaf blades, ¼ to ⅓ inch broad, and the sheaths have long stiff hairs. The flowering part is made up of 3 to 11 slender, fingerlike branches, 2 to 6 inches long, which may all arise from the same point at the stem tip (as in Bermudagrass), but usually several branches in addition arise a short distance below the tip. The small spikelets, ⅛ to 3/16 inch long, lie very close to the branch stems and come from only one side of the axis. The light yellow oval grain is about 1/12 inch long.

DISTRIBUTION — Large crabgrass, introduced from Europe, is a weed of moist soil. It is particularly obnoxious in lawns, where it forms coarse basal rosettes of leaves, but also is very common in cultivated fields, along streams, ditch banks, roadsides, and washes in southern and central Arizona, uncommon northward except in certain local areas; 100 to 6,000 feet elevation; flowering June to October.

### SMOOTH CRABGRASS, small crabgrass — *Digitaria ischaemum* (Schreb.) Muhl.

DESCRIPTION — Similar to crabgrass, but smaller and not hairy. The bract enclosing the grain is blackish brown rather than pale yellow. A "recent" arrival in the Bermuda lawns of southern Arizona, and spreading rapidly. On the University of Arizona campus, smooth crabgrass was found in nearly pure stands occupying areas up to 30 feet or more in diameter.

Fig. 14. Large crabgrass *(Digitaria sanguinalis)*. Plants with fingerlike inflorescences. *a.* Spikelet. *b.* Spikelet side view. *c., d.* Two views of grain or floret.

# DESERT SALTGRASS, inland saltgrass, seashore saltgrass, marsh spikegrass, saltgrass

### GRASS FAMILY — Gramineae

## DESERT SALTGRASS — *Distichlis stricta* (Torr.) Rydb.

DESCRIPTION — Desert saltgrass is a low stiff perennial, 4 to 16 inches high, reproducing by seed and by tough, extensively creeping underground stems. These root at the joints and produce new stems, often forming dense colonies. The lower part of the hard rigid aboveground stems usually lies on the ground, then becomes erect.

The narrow leaves are alternate, but come off the stem in two rows. They are ½ to 4 inches long, and about ⅛ inch broad, sometimes folded lengthwise, or the edges rolled inward. The flowering part of the plant at the top of the stem is yellowish, short, and narrow. Although all the plants look alike, including their spikelets, there are actually two kinds of saltgrass plants, male and female. Each spikelet contains 5 to 10 bractlike flowers. Those on the male plants bear stamens only, and the female only pistils. The grain remains enclosed in a pointed greenish or purplish hull.

DISTRIBUTION — A native weed in moist to dryish mostly alkaline soil, or sometimes on sandy or heavy bottomland soil. Common on the Little Colorado and Salt River Valley drainage flood plains, and marshes in the desert and northern desert, in Apache, Navajo, Coconino counties and southward; 100 to 6,000 (rarely higher) feet elevation; flowering from May to October. From its native habitat, saltgrass has spread to the irrigated lands, and become a pest in ditches, cotton fields, and other crops, as in the Yuma and Moencopi (Navajo County) areas.

Fig. 15. Desert saltgrass *(Distichlis stricta)*. Flowering plant with thickened rhizomes. Also lower part of an erect stem. *a.* Spikelet. *b.* Single floret.

# BARNYARDGRASS

GRASS FAMILY — Gramineae

## BARNYARDGRASS — *Echinochloa crusgalli* (L.) Beauv.

DESCRIPTION — A stout summer annual, ½ to over 4 feet high, which reproduces only by seed. It is often spreading and prostrate at the base, rooting at the lower nodes, sometimes forming large clumps.

The hairless leaves are dense, with blades 4 to 20 inches long, and ¼ to ¾ inch broad. The flowering tops are 3 to 8 or 10 inches long, with the erect branches 1 to 2¼ inches long.

The green or purplish spikelets, about ⅛ inch long (excluding the bristle when present), are densely and irregularly crowded on the branches, and almost without stalks. They typically are stiff hairy and short awned to sharp pointed at the tip, but sometimes ending in a bristle up to 1½ inches long.

The pale yellow, shiny grain is flat on one side and round on the other. A single plant has been estimated to produce as many as 40,000 grains.

DISTRIBUTION — Barnyardgrass is a European introduction that has become a most troublesome weed in moist soil in all agricultural areas in the state. It is common in irrigated fields, orchards, pastures, roadside swales, reservoirs, ditches, and streams throughout most of Arizona; 100 to 7,000 feet elevation; flowering June to October. The name "crabgrass" should not be applied to this grass since that is the accepted name for *Digitaria* spp.

Fig. 16.  Barnyardgrass *(Echinochloa crusgalli).*  Flowering plant.
*a.* Spikelet spread open.  *b.* Two spikelets with short and long awn.  *c.* Grain.

## JUNGLERICE, watergrass
### GRASS FAMILY — Gramineae

### JUNGLERICE — *Echinochloa colonum* (L.) Link.

DESCRIPTION — A spreading annual, ⅔ to 2 or 3 feet high, which reproduces by seeds. Junglerice closely resembles barnyardgrass; the stems are spreading and prostrate, often rooting at the base, then ascending. The first growth form frequently is a dense rosette of leaves at the ground level. These and the other leaves often have similar purple bands, but are narrower, ¼ inch broad or less. Barnyardgrass differs principally in the flowers: the flowering part is shorter, 2 to 6 inches long, and the branches are shorter, ⅜ to ¼ (or to 1) inch long. Also none of the spikelets end in a bristle, but are merely sharp pointed. They are crowded on the stem in 2 to 4 regular rows, rather than being irregularly arranged.

DISTRIBUTION — Junglerice, more widespread than barnyardgrass, is a pest in central and southern Arizona's irrigated lands. Found in such crops as sorghum, cotton, alfalfa, lettuce, and melons. It is also annoying in lawns (where it forms thick basal rosettes), flowerbeds, and gardens; 150 to 5,500 feet elevation; flowering from February to October or November.

Seeds of both junglerice and barnyardgrass are used in Asia and Africa for human food.

42

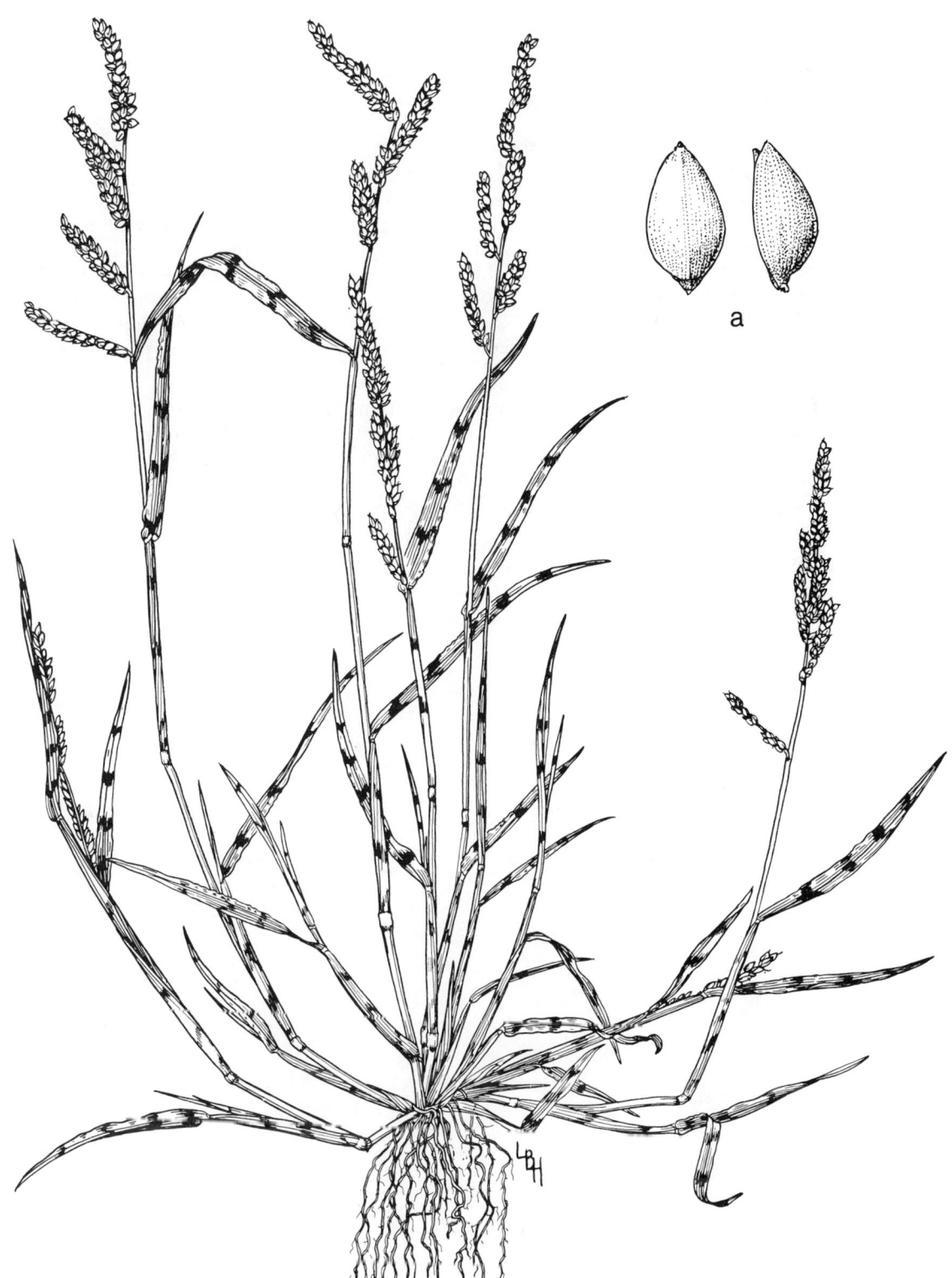

Fig. 17. Junglerice *(Echinochloa colonum).* Flowering plant,
with purple bands on the leaves. *a.* Top with side view of grain.

43

# STINKGRASS, strong-scented lovegrass
GRASS FAMILY — Gramineae

## STINKGRASS — *Eragrostis cilianensis* (All.) Link

DESCRIPTION — Stinkgrass is an annual, branching from the base, ⅓ to 2 feet long, which reproduces only by seed. The stems are often prostrate then ascending, with minute glandular pits in a ring below the joints. These pits may also be scattered on the leaves, flower stalks, and the margins of the flowering bracts, supposedly giving off a cockroachlike odor that is offensive to livestock. The flowering part of the plant is 2 to 10 inches long, and widely spreading. The spikelets are densely crowded, ¼ to ¾ inches long, with 12 to 40 flowers or florets. The reddish brown oval grains are about 1/16 inch long, and pointed at both ends.

DISTRIBUTION — A very common weed from early summer through fall in cultivated crops such as sorghum, cotton, citrus, and alfalfa; also in gardens, roadsides and waste places, or in heavy soiled drainage lands throughout most of the state except the northeast part; from 100 to 6,000 feet elevation; flowering from May to October. Seeds of this European weed often occur as an impurity in small commercial seeds.

POISONOUS PROPERTIES — Horses are reported to have been poisoned by eating large quantities of the fresh or dried plant over a period of time, but none in Arizona.

Fig. 18. Stinkgrass *(Eragrostis cilianensis)*. Flowering plant.
*a.* Portion of stem showing ring of glands below the joint. *b.* Spikelet. *c.* Grain.

45

# SOUTHWESTERN CUPGRASS

GRASS FAMILY — Gramineae

## SOUTHWESTERN CUPGRASS — *Eriochloa gracilis* (Fourn.) Hitchc.

DESCRIPTION — A spreading annual, 1 to 3 feet high, which reproduces by seed and by the prostrate stems rooting at the nodes. Without the flowering tops, southwestern cupgrass is very similar to the crabgrasses. It may be distinguished from large crabgrass, however, by the fact that the bright green leaves are not hairy. The flowering part is hairy, 2 to 6 inches long, with short, erect to spreading branches 1 to 2 inches long. The spikelets are about 3/16 inch long, tapering into a point at the tip. Each spikelet has a cuplike, sometimes darkened ring around the base, which distinguishes the cupgrasses from most other grasses. The yellowish oval grain is flat on one side and rounded on the other, about ⅓ inch long with a short sharp point at the tip.

DISTRIBUTION — Southwestern cupgrass is a native, and one of the most common and troublesome weeds in summer lawns. It is very commonly mistaken for crabgrass. It is also abundant along roadsides, city streets, cultivated fields, streams, washes, canyons, ditches, and pools throughout southern Arizona; 125 to 5,500 feet elevation; flowering June to October.

## CANYON CUPGRASS — *Eriochloa lemmoni* Vasey & Scribn.

DESCRIPTION —An annual ⅔ to 2 or 3 feet high, similar to southwestern cupgrass, but the leaf blades are hairy. Frequently found on disturbed soil of roadsides, canyons, and washes in Cochise, Pima, and Santa Cruz counties; 2,400 to 5,700 feet elevation; flowering August to October.

Fig. 19. Canyon cupgrass *(Eriochloa lemmoni)*. Flowering plant.
*a.* Spikelet. *b.* Grain. *c.* Southwestern cupgrass *(E. gracilis)*. Spikelet.

47

## WILD BARLEY, common foxtail
### GRASS FAMILY — Gramineae

### WILD BARLEY — *Hordeum leporinum* Link

DESCRIPTION — A many branched, spreading, or nearly prostrate annual, 6 to 20 inches high, which reproduces by seed. The broad, flat leaf blades are 1½ to 4 inches long. The thick erect flowering spike is 2 to 3 inches long, and usually partially enclosed by the uppermost expanded leaf sheath. It breaks apart when it is mature. The stout, stiff bristles are ¾ to 1½ inches long. The yellow grain is about ¼ inch long and hairy at the top.

DISTRIBUTION — Introduced from Europe, wild barley is a weed pest starting early in the spring in cultivated crops, especially grain and alfalfa fields. It is common on disturbed soil of roadsides, irrigation ditches, vacant lots, and lawns throughout Arizona, except the northeastern part; 100 to 9,000 feet elevation; flowering mostly in March and April at lower altitudes, and to October at higher elevations.

### FOXTAIL BARLEY — *Hordeum jubatum* L.

DESCRIPTION — Foxtail barley is a tufted perennial, often appearing annual, 1 to 2 feet high, reproducing by seeds and by inconspicuous rhizomes. It is similar to wild barley, but larger; the flowering spikes, 2 to 4 inches long, are nodding, not erect; the pale green or reddish bristles are much longer, ¾ to 3⅛ inches long, and finer, not as stiff. The flower heads are only about 1 inch broad until maturity, when they spread and are very bushy, as in squirreltail (*Sitanion hystrix*), except the bristles are much finer and not stiff.

DISTRIBUTION — A native weed of moist soil on disturbed ground along streams, lakes, roadsides, and in irrigated pastures and ditches (but not usually in cultivated fields). At maturity, it is sometimes injurious to stock because the bristles and sharp joints pierce their mouths, nostrils, and skin. Found in Apache, Navajo, Coconino, and Maricopa counties; 5,000 to 7,500 feet elevation; flowering June to September.

Fig. 20. Wild barley *(Hordeum leporinum)*. Flowering plant.
*a.* Spikelets. *b.* Floret with long bristle.

# RED SPRANGLETOP

GRASS FAMILY — Gramineae

## RED SPRANGLETOP — *Leptochloa filiformis* (Lam.) Beauv.

DESCRIPTION — Tall or spreading weak annuals, 4 inches to 4 feet high, often reddish or purplish. The leaf blades are thin and flat, and the sheaths bear long slender hairs. The mature flowering tops are extremely variable in length, but are about ⅓ to ½ the length of the stems, and may be 1½ feet long. The very slender branches are 1 to 5 inches long, and spreading at maturity. The tiny spikelets, less than ⅛ inch long, are not awned.

DISTRIBUTION — Red sprangletop is a common weed of roadsides, citrus orchards, ditchbanks, cotton, sorghum, and alfalfa fields, gardens, sandy washes, and dry slopes throughout Arizona, except Navajo and Mohave counties; 100 to 5,000 feet elevation; flowering May to October or November.

## BEARDED SPRANGLETOP — *Leptochloa fascicularis* (Lam.) Gray

DESCRIPTION — Coarse, succulent annual, forming rather large clumps or sometimes dwarfs, and only 4 to 6 inches high. The leaf blades are narrow and flat, or inrolled, tubelike, and a bluish-green color.

DISTRIBUTION — Bearded sprangletop is a weed of wet places and alkaline soil, in irrigated crops, along ditches, streams, reservoirs, roadside swales, and brackish water along lakes and rivers in Navajo, Graham, Gila, Pinal, Cochise, and Pima counties; 1,500 to 5,000 feet elevation; flowering May to October or November.

50

Fig. 21. Red sprangletop *(Leptochloa filiformis)*. Plant and flowering branch.
*a.* Spikelet. *b.* Grain. *c.* Bearded sprangletop *(L. fascicularis)*. Spikelet.
*d.* Grain of bearded sprangletop.

51

# MEXICAN SPRANGLETOP

GRASS FAMILY — Gramineae

## MEXICAN SPRANGLETOP — *Leptochloa uninervia* (Presl.) Hitchc. & Chase

DESCRIPTION — Very similar to bearded sprangletop, but the mature spikelets are no more than ¼ inch long, and lead colored. The spikelets of bearded sprangletop are ¼ to ½ inch long, and not lead-colored. Mexican sprangletop grows in the same type of wet places as bearded sprangletop, and is found in Mohave, Pinal, Maricopa, Yuma, and Pima counties; 100 to 3,500 feet elevation; flowering June to October.

This weed is particularly troublesome in the fall in winter barley fields, since it germinates sooner than the barley grains. In the spring it is a great pest in cantaloup crops.

Fig. 22. Mexican sprangletop *(Leptochloa uninervia)*.
Plant and flowering branch. *a.* Spikelet. *b.* Grain.

# WITCHGRASS, tumble panic, witches hair
GRASS FAMILY — Gramineae

## WITCHGRASS — *Panicum capillare* L.

DESCRIPTION — A bushy, conspicuously hairy annual with a shallow root which reproduces only by seeds. The stems are hairy, erect or mostly spreading, and branched from the base, ⅓ to 3 feet high. The leaves are ¼ to ½ inch broad, and are covered (especially the bases) by long, soft or stiff hairs. The mature flowering part is very large and bushy, often half the length of the entire plant, greatly spreading and diffusely branched, with the branches stiff and threadlike. It is brittle when ripe, and often the entire flowering part breaks away and is blown about by the wind as a "tumbleweed." One spikelet occurs at the tip of each little branch. The spikelets are ½ to ⅛ inch long, with a smooth and shiny yellow or gray grain, which is about 1/16 inch long.

DISTRIBUTION — Witchgrass is a native weed of dryish soil. Common along sandy canyon washes, streams, open ground and pasturelands, it has increased and spread in Arizona with the increase in agriculture. It has become abundant in the irrigated lands, cotton and alfalfa fields, and also occurs in flower beds, gardens, and waste places throughout most of the state; 100 to 8,000 feet elevation; flowering from May at lower elevations, or from July at higher elevations, to October. The grains are frequently found in commercial alfalfa seed and other small seeds.

Fig. 23. Witchgrass *(Panicum capillare)*. Plant and
flowering stem. *a.* Spikelet. *b.* Two views of grain.

55

# BROWNTOP PANICUM

GRASS FAMILY — Gramineae

### BROWNTOP PANICUM — *Panicum fasciculatum* Sw. var. *reticulatum* (Torr.) Beal

DESCRIPTION — A tall annual, erect or spreading, and branching from the lower stem joints, 1 to nearly 4 feet high, reproducing by seeds. The leaves are thinly hairy or hairless, and ¼ to ½ inch broad. The flowering part is short, mostly 2 to 6 inches long, with erect to slightly spreading spinelike branches, 2 to 4 inches long.

The spikelets are green when young, and yellow or bronze when ripe. They are hairless, less than ⅛ inch long, and minutely cross ridged. The grain is dull, also minutely cross ridged and ⅛ inch or less long, with a blunt tip.

DISTRIBUTION — Browntop panicum is a native weed, infrequent in Arizona in sandy washes, river bottoms, and waste places until a few years ago. It has now become common, and is often a nuisance in cultivated crops in the irrigated areas of southern Arizona, such as the Salt and Gila rivers, Santa Cruz, Avra, and Yuma valleys; 100 to 3,500 feet elevation; flowering June to October.

### ARIZONA PANICUM — *Panicum arizonicum* Scribn. & Merr.

DESCRIPTION — Closely resembles browntop panicum, but is shorter, ½ to 2 feet high. The spikelets are about the same color, but are short hairy, slightly longer than ⅛ inch, and not minutely cross ridged. The grains are short pointed.

DISTRIBUTION — Arizona panicum is a native weed of mesas, sandy washes, rocky slopes, and canyons throughout southern Arizona; up to 5,000 feet. It also is a common farm weed found in the same general places, and flowering at the same time as browntop panicum.

Fig. 24. Browntop panicum *(Panicum fasciculatum* var. *reticulatum)*. Habit of plant and flowering branch. *a.* Lower and upper view of spikelet. *b.* Two views of grain. *c.* Arizona panicum *(P. arizonicum)*. Spikelet. *d.* Grain.

# DALLISGRASS

GRASS FAMILY — Gramineae

## DALLISGRASS — *Paspalum dilatatum* Poir.

DESCRIPTION — Dallisgrass is a tall perennial from a hard, knotty base, 1½ to 5 feet high, which reproduces only by seed. It has flat leaf blades about 2 to 6 inches long. The flowering part consists of 3 to 5 very narrow branches 1 to 3 inches long, alternate along the upper part of the stem. These bear the spikelets on one side only.

The small pointed spikelets, about ⅛ inch long, are stalkless, and lie in even rows close to the branch. The dark purple stigmas are often conspicous. The smooth shiny grain is yellowish, nearly circular, and slightly less than ⅛ inch long.

DISTRIBUTION — Dallisgrass is a South American species that grows in the moist or marshy soils of grazing pastures, not on the rangelands. As a weed it is troublesome in lawns, but also occurs in alfalfa and other crops in southern Arizona — particularly in Pinal, Maricopa, and Pima counties. This grass is planted as a permanent pasture grass in southern Arizona, and has been grown in Apache County, but less successfully. It flowers from April to November.

POISONOUS PROPERTIES — Although dallisgrass is good forage, it is very susceptible to the fungus attacks of paspalum ergot (*Claviceps paspali* Stevens & Hall). The fungus produces a dark sticky substance similar to axle grease, known as "honey dew," which covers the grass flowers. Paspalum ergot is the chief cause of ergot poisoning reported in Arizona, principally of cattle running in pastures containing dallisgrass.

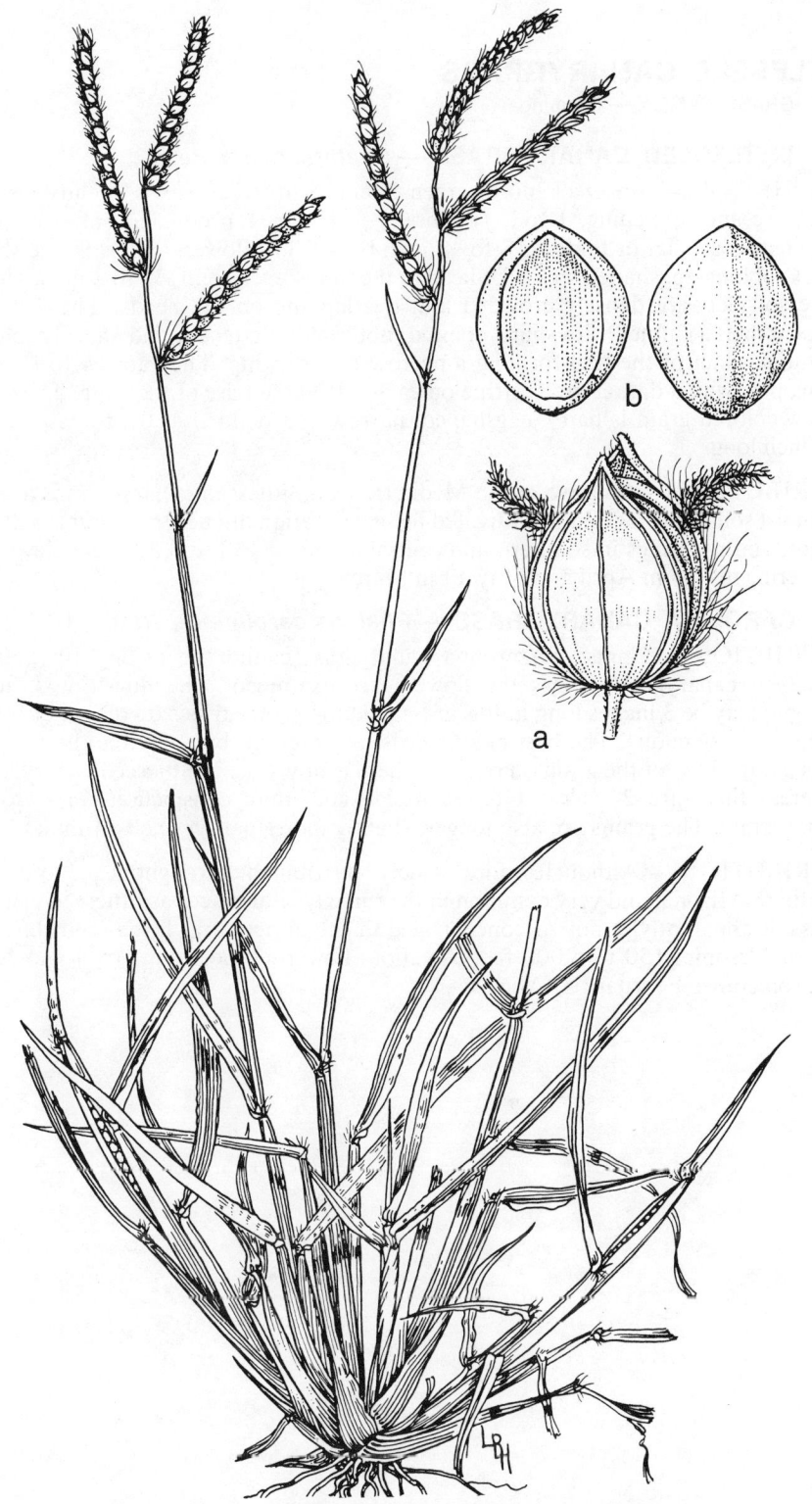

Fig. 25. Dallisgrass *(Paspalum dilatatum)*. Flowering plant.
*a*. Spikelet, showing protruded stigmas of flower. *b*. Two views of grain.

59

# LITTLESEED CANARYGRASS
GRASS FAMILY — Gramineae

### LITTLESEED CANARYGRASS — *Phalaris minor* Retz.

DESCRIPTION — An erect bluish green annual with weak stems slightly bent at the base and branching, 1 to 2 (or more) feet high, reproducing only by seeds. The leaves are flat or folded, ⅛ to ⅜ inch broad. The flowering part is one short, thick, oblong-eggshaped spikelike head at the top of each stem, ¾ to 2 inches long. The spikelets are densely crowded and overlapping on the heads. The 2 broad outer bracts (glumes) are sharp pointed, noticeably flattened, and sharply folded, the upper half of the fold forming a narrow papery wing. They are ⅛ to ¼ inch long, pale with a darker green stripe on each side at the base of the wing. The shiny straw colored grain is hairy, eggshaped, narrow and pointed at the tip, and about ⅛ inch long.

DISTRIBUTION — A native of the Mediterranean, littleseed canarygrass is a weed in moist soil in the margin of cultivated fields, irrigation ditches, reservoirs, bottom-lands, and roadsides in southern and central Arizona; 150 to 3,500 feet elevation; flowering mostly in April and May, or in March.

### CAROLINA CANARYGRASS — *Phalaris caroliniana* Walt.

DESCRIPTION — Plants of Carolina canarygrass cannot be distinguished from littleseed canarygrass unless the flowers are examined. The flowering head is oblong, may be 3 inches long in this grass, and the green stripes on the outer bracts are not conspicuous. The best character is in the grain, but requires the aid of a lens. At the base of the grain on one side there is tiny scale in littleseed canarygrass, whereas there are 2 scales, 1 on each side and more conspicuous in Carolina canarygrass. The grains are also longer, slightly more than ⅛ inch (4 mm.) long.

DISTRIBUTION — Although more widely distributed throughout central and southern Arizona, and very common in the same type of places as littleseed canary-grass, it apparently is not so concentrated in the agricultural fields, and thus not as troublesome; 150 to 6,000 feet elevation; flowering mostly in April and May, but sometimes February to August.

Fig. 26. Littleseed canarygrass *(Phalaris minor).* Plant with terminal spikelike, flower heads.
a. Spikelet. b. Grain. c. Carolina canarygrass *(P. caroliniana).* Flowering branch.
d. Spikelet, showing two glumes and grain.

## ANNUAL BLUEGRASS, walkgrass
GRASS FAMILY — Gramineae

### ANNUAL BLUEGRASS — *Poa annua* L.

DESCRIPTION — A low spreading annual which reproduces only by seeds. The tufted stems, often bent and rooting at the base, are 4 to 12 inches high. The bright green leaf blades are about 3 inches long. The flowering part is short, 1 to 4 inches long, with spreading branches and few spikelets. The bright amber grain is pointed at both ends.

DISTRIBUTION — Annual bluegrass is found in many moist places, but it primarily a nuisance in lawns where it may grow very luxuriantly, covering large areas for a very short period in the winter and spring, then dies out, leaving ugly patches. This European introduction occurs throughout most of the state; up to 8,000 feet elevation; flowering from March to September (higher altitudes).

Fig. 27. Annual bluegrass *(Poa annua)*. Flowering plant. *a*. Spikelet. *b*. Grain.

# RABBITFOOTGRASS

GRASS FAMILY — Gramineae

### RABBITFOOTGRASS — *Polypogon monspeliensis* (L.) Desf.

DESCRIPTION — A yellowish green annual, ⅓ to 2 feet high, with the stems often sharply bent and spreading at the lower joints, reproducing only by seeds. The flowering part is a soft silky spike, 1 to 4 (sometimes 6) inches long, with delicate glistening, yellowish or tawny bristles. The flowering top is so dense it appears to be unbranched, but the innumerable tiny spikelets, about 1/12 inch long, are actually crowded on little short branches.

The 2 outer bracts in each spikelet end in a slender bristle about ¼ to ⅜ inch long. These give the flower head its bristly appearance. The plump grain is amber colored, about 1/12 inch long.

DISTRIBUTION — A European weed, rabbitfootgrass is found in moist soil throughout most of the state in irrigated areas, cultivated fields, pastures, and ditches. It also occurs in river bottoms, swales, along roadsides, streams, and in mountain canyons; 100 to 8,200 feet elevation; flowering March to October.

### DITCH POLYPOGON — *Polypogon interruptus* H.B.K.

DESCRIPTION — A tufted perennial, but very similar to rabbitfootgrass, and sometimes hard to distinguish. The flowering part is more spreading, with the branches obvious. It is much less spikelike in appearance. The bristles are shorter, about ⅛ inch long.

Ditch polypogon, also introduced from Europe, is found in the same habitats as rabbitfootgrass, but is not as common; 150 to 7,500 feet elevation; flowering May to October.

Fig. 28. Rabbitfootgrass *(Polypogon monspeliensis)*. Flowering plant.
*a*. Spikelet. *b*. Ditch polypogon *(P. interruptus)*. Spikelet.

# MEDITERRANEANGRASS

GRASS FAMILY — Gramineae

## MEDITERRANEANGRASS — *Schismus barbatus* (L.) Thell.

DESCRIPTION — A low tufted annual, erect, spreading, or often forming large prostrate mats on the ground, reproducing by seeds only. There are many weak stems 2 to 14 inches in length, with very narrow leaf blades, 1/12 inch or less broad, and 4 inches or less long. The flowering part is a small cluster of short purplish branches grouped close together on the upper part of the stem, ½ to 2½ inches long. The spikelets, often purple tinged, are ¼ to ⅜ inch long, with 2 long outer bracts (glumes) ¾ to as long as the rest of the spikelet. The shiny translucent grain is oval to eggshaped, and only about 1/25 inch long.

DISTRIBUTION — Introduced from the Mediterranean region, this grass has spread rapidly, and is now very common in vacant lots, city streets and roadsides, irrigated pastures and cultivated fields, also on dry slopes, desert mesas, river bottoms or plains in southern Arizona; 100 to 3,700 feet elevation; flowering January to May. Locally abundant on some southwestern ranges in the state, it assumes some importance as a spring forage plant, although it is a relatively recent arrival.

## ARABIANGRASS — *Schismus arabicus* Nees.

DESCRIPTION — This rapidly spreading grass, introduced from Asia or Africa, looks exactly like Mediterraneangrass. It differs in that the flowering bracts of the spikelets are more deeply notched at the tip, and are long hairy on the back. The 2 weeds apparently intergrade freely, and cannot always be clearly differentiated. Arabiangrass is rapidly appearing throughout the Mediterraneangrass range, and is very common in some localities.

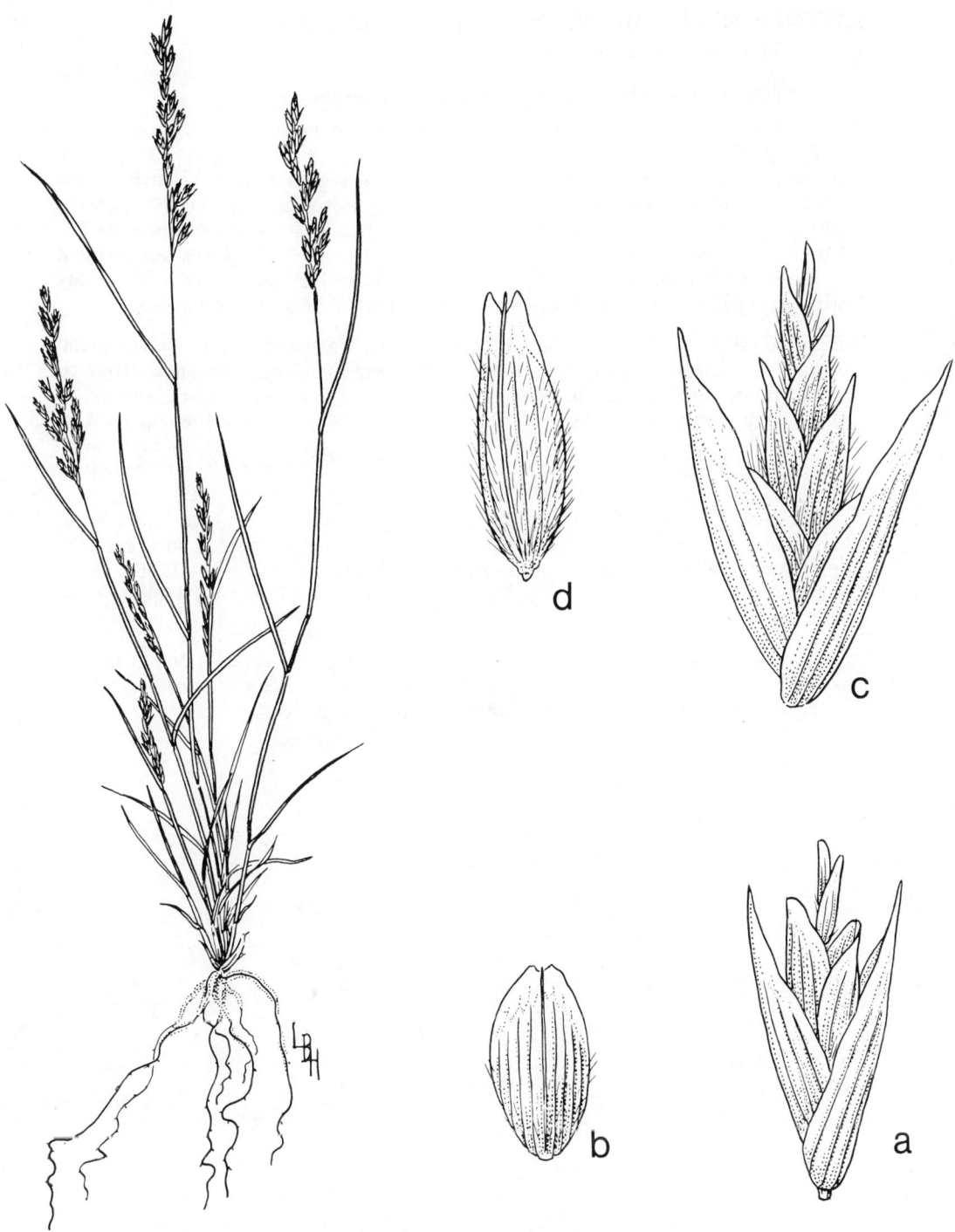

Fig. 29. Mediterraneangrass *(Schismus barbatus).* Flowering plant. *a.* Spikelet.
*b.* Floret. *c.* Arabiangrass *(S. arabicus).* Spikelet. *d.* Floret of arabiangrass.

## GREEN FOXTAIL, green bristlegrass, bottlegrass
### GRASS FAMILY — Gramineae

### GREEN FOXTAIL — *Setaria viridis* (L.) Beauv.

DESCRIPTION — Weak annuals forming spreading clumps, ½ to 1¾ or 3 feet high, reproducing only by seeds. The flattened leaf blades are usually less than 6 inches long, and ¼ to ⅜ inch broad. The flowering part is a bristly cylindrical spike at the end of a stem ¾ to 3 inches long. The spikelets, about 1/12 to ⅛ inch long, are densely crowded on the spike. At the base of each, there arises 1 to 3 (sometimes 4) tawny or purplish bristles, mostly 3 to 4 times longer than the spikelets. The nearly oval grains are about 1/16 inch long, greenish to dark brown and faintly wrinkled, flattened on one side and rounded on the other.

DISTRIBUTION — Green foxtail, native of Europe, is a common weed in moist soil throughout Arizona, and is a nuisance in cultivated fields in irrigated valleys. It is also found in lawns, ditches, along roadsides, streams, and in barren spots in pine forest openings; 100 to 8,200 feet elevation; flowering June to November.

### YELLOW FOXTAIL, YELLOW BRISTLEGRASS — *Setaria glauca* (L.) Beauv. (*S. lutescens* [Weigel] Hubbard)

DESCRIPTION — An annual, closely resembling green foxtail. The leaves are longer, 4 to 12 inches long, and the flowering spikes may be 4 inches long. It differs principally in that there are at least 5, and usually more (5 to 20), bristles at the base of each spikelet. The grains are about 1/12 inch long, and thick pointed at the tip.

DISTRIBUTION — Yellow foxtail, also introduced from Europe, is found in the same type of places with the same general distribution as green bristlegrass, but is infrequent; 100 to 7,500 feet elevation; flowering July to October.

Fig. 30. Green foxtail *(Setaria viridis)*. Flowering plant. *a.* Two views of spikelet and two of grain. *b.* Yellow bristlegrass *(S. glauca)*. Two views of grain and two of spikelet.

# SQUIRRELTAIL

GRASS FAMILY — Gramineae

## SQUIRRELTAIL — *Sitanion hystrix* (Nutt.) J. G. Smith

DESCRIPTION — Squirreltail is a tufted perennial, ½ to 1⅔ feet high, reproducing only by seeds. The leaves are soft hairy to nearly hairless, and 1/12 to ⅕ inch broad. The flowering part is a dense stiff spike, bushy from the many long slender bristles. It is ¾ to 3 (or 4) inches long, and breaks apart easily at each place where the spikelets are attached to the stem. The flowering bracts of the spikelets all end in long, barb-margined spreading bristles, ½ to 3½ inches long. Since there are many spikelets, and each produces a variable number (about 7 or more) of these long widely spreading bristles, the flower spike is conspicuously bushy at maturity.

DISTRIBUTION — Squirreltail is a native weed. It is abundant throughout the state, and in many types of habitats. A troublesome weed of roadsides, yards, and waste places, also on barren places of dry hillsides, rocky slopes, mountain meadows, open forests, to above timberline; 2,400 to 11,500 feet elevation; flowering March to September. It is fair forage when young, but the mature awns work into the flesh and wool of the animals, causing inflammation.

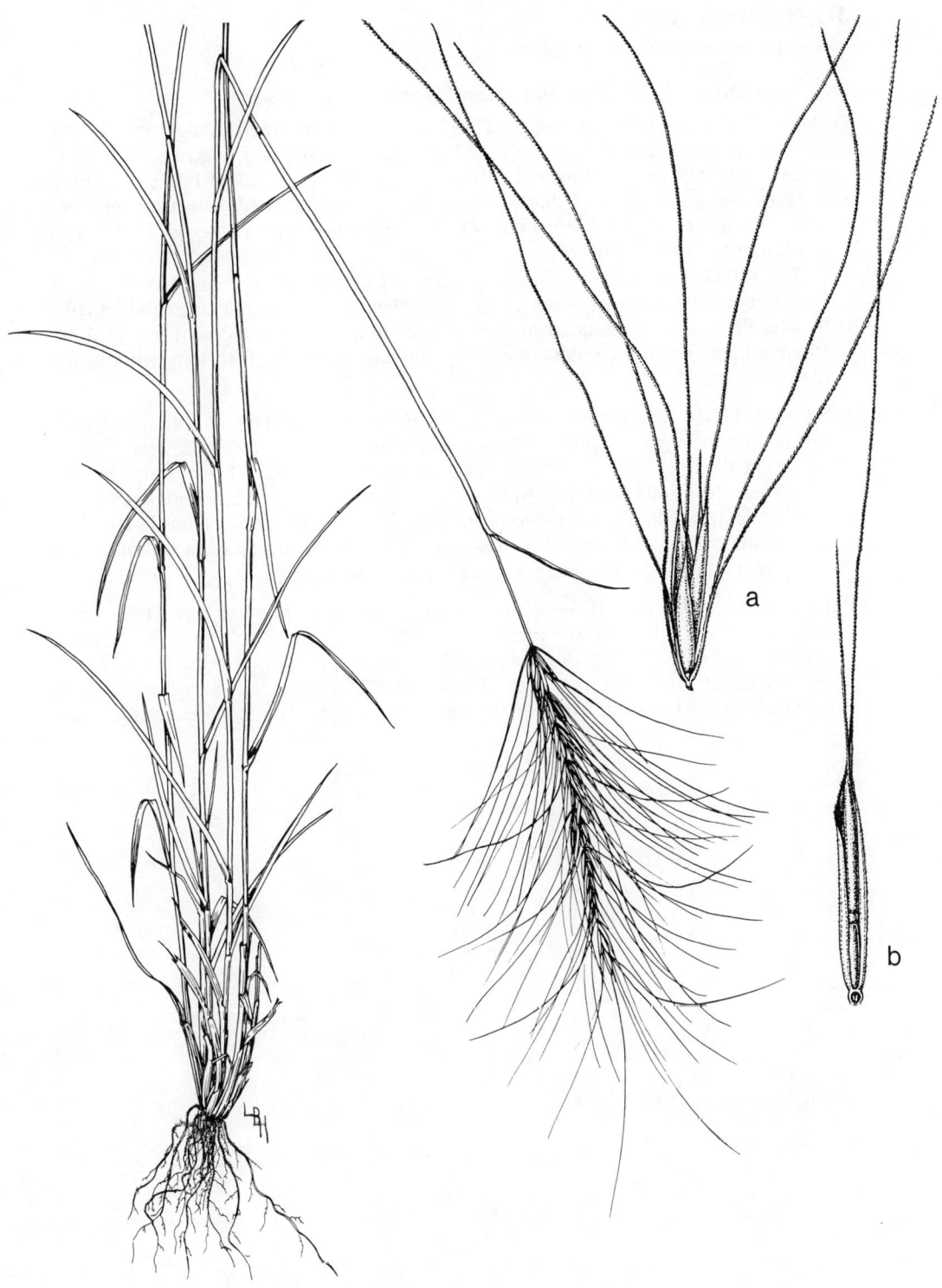

Fig. 31.  Squirreltail *(Sitanion hystrix).* Flowering plant.
*a.* Spikelet. *b.* Floret.

# JOHNSONGRASS

GRASS FAMILY — Gramineae

## JOHNSONGRASS — *Sorghum halepense* (L.) Pers.

DESCRIPTION — Johnsongrass is a prohibited noxious weed in Arizona. It is a coarse perennial, 3 to 7 feet high, and very leafy, spreading by seeds and by an extensive system of underground rhizomes. The bright green leaf blades are up to 2 feet long, and ¼ to ¾ inch broad. The many-branched flowering tops are loose, open, and ½ to 2 feet long. The drooping branches come off 2 or 3 at a joint, and are naked below.

The spikelets occur in pairs, but in threes at the tips of the branches, 1 (or 2) is stalked and bears stamens only, while the other is stalkless, thickened, and fertile. The fertile spikelet is about ¼ inch long, and has a twisted, once-bent bristle about ½ inch long. The dark reddish brown grains are nearly ⅛ inch long without the hull.

DISTRIBUTION — Johnsongrass is one of the most abundant and vicious weeds throughout the state, serious in all important summer crops. It may be found along irrigation ditches, cultivated fields, and moist waste places of any type; 100 to 6,000 feet elevation; flowering April to November. This weed can completely take over agricultural lands unless constant control methods are practiced. Its stout underground stems may be 2½ feet deep, and the grains may lie dormant for many years, making complete eradication almost impossible.

POISONOUS PROPERTIES — Johnsongrass ordinarily is good feed, but sometimes the plant, particularly the leaves, contain hydrocyanic (prussic) acid, a cyanide type of poisoning. Any factor which interrupts normal growth may cause the release of HCN within plants. Rapid growth of new leaves, wilting due to drought, frost, freezing, cutting, or trampling are the most dangerous events.

Fig. 32. Johnsongrass *(Sorghum halepense)*. Plant with stout rhizomes, also flowering branch. *a.* Group of three spikelets from tip of stem (two sterile and stalked, the third fertile and awned). *b.* Grain with hull.

73

# YELLOW NUTSEDGE, yellow nutgrass

SEDGE FAMILY — Cyperaceae

## YELLOW NUTSEDGE — *Cyperus esculentus* L.

The nutsedges are prohibited noxious weeds in Arizona.

DESCRIPTION — A tough erect perennial, 1 to 3 feet high, reproducing by seeds and by many deep, very slender rhizomes which form weak runners above the ground, and small tubers or nutlets at the tips of underground stems. The tubers are dark, unevenly globeshaped, ⅜ to ¾ inch long, and edible, tasting somewhat like almonds. Near the base of the triangular stem, a cluster of 3-ranked, grasslike leaves arises which are often longer than the stem, and ⅛ to ⅓ inch broad. The umbrellalike flowering tops have a few threadlike stems of different lengths radiating out like spokes from the stem tip. These have numerous yellowish to golden brown spikelets on the upper part. The spikelets are very narrow, flattened, 1/16 to 1/12 inch broad, and ¼ to 1 (or 1½ ) inch long. The leaves beneath the flowering tops are 2 to 10 inches or more long, and some are much longer than the flowering stems. The brownish 3-angled achene, about 1/16 inch long, is widest at the top.

DISTRIBUTION — Yellow nutsedge, an Old World introduction, is a noxious weed of wet soil. The nutsedges are the most difficult weeds to control in cultivated fields, often forming a solid cover over large areas in cotton fields, sorghum and alfalfa pastures, flood plains, dams, ditches, and along streams and roadsides. Yellow nutsedge is common throughout most of the state; 100 to 8,200 feet elevation; flowering May to November or fall frosts.

## PURPLE NUTSEDGE, purple nutgrass — *Cyperus rotundus* L.

DESCRIPTION — Purple nutsedge closely resembles yellow nutsedge, but the mature stems are usually longer than the basal leaves. The leaves below the flowering heads, 1 to 5 inches long, are about the same length as the flower stems, few are longer. The spikelets are dark brown-purple, and the runners are fewer, woody, and thicker. The nutlets are oblong and covered by persistent reddish scales, whereas they are almost smooth in yellow nutsedge at maturity, and unevenly globose.

DISTRIBUTION — Purple nutsedge is largely confiined to the valleys of southern Arizona; 100 to 4,000 feet elevation; flowering from May to fall frosts. In all summer crops, in lawns, ditchbanks, and field borders; the rhizomes, runners, and tubers make the nutsedges almost impossible to eradicate.

Fig. 33. Purple nutsedge *(Cyperus rotundus)*. Flowering plant showing some of the underground stems with nutlets or tubers at the tips. *a*. Spikelet in flower. *b*. Achene.

# ROCKY MOUNTAIN IRIS

IRIS FAMILY — Iridaceae

## ROCKY MOUNTAIN IRIS — *Iris missouriensis* Nutt.

DESCRIPTION — A perennial, 1 to 2½ feet high, reproducing from seeds and from dark reddish thick woody rhizomes, mostly horizontal and branching, bearing many stout roots. The long narrow bluish green leaves in clusters at the base of the flower stems are 2-ranked, flat or folded lengthwise, and enfolding one another, ⅓ to 2 feet long, and ⅛ to ⅜ inch broad.

The flowering stems are 1 to 2½ feet high, usually leafless or with an occasional small leaf, bearing 2 to several flowers, one flower blossoming after another. The large sweet smelling flowers are pale blue or violet, about 3 inches in diameter, and 3 to 4 inches long. The base of the flowers is enclosed by a pair of bracts, leaflike at first, becoming thin and dryish, 1½ to 3 inches long. The 3 outer parts of the flower curve downward, and are longer than the 3 inner erect petals. The 3 branches of the style are large, colored similar to the petals, and spread outwardly covering the 3 stamens.

The large oblong seedpods, 1 to 1½ inches long, ½ to ¾ inch thick, are 3-celled, with many seeds, usually in 2 rows in each cell. The dark reddish brown seeds are somewhat wedgeshaped, with irregular depressions and ridges, or D-shaped and plump, ⅛ to ¼ inch long.

DISTRIBUTION — Rocky Mountain iris is a native range weed of wet soil and high elevations. It is common on the high plateaus, open mountain meadows, near springs, in wet barren flats, or other places more or less denuded of vegetation. It becomes more common on overgrazed, deteriorated ranges. Found in Navajo, Apache, and Coconino counties, and in the mountains of southeastern Arizona; 6,000 to 9,500 feet elevation; flowering May to September.

Fig. 34. Rocky mountain iris *(Iris missouriensis).* Flowering
plant and branch with two seedpods. *a.* Seed.

77

# GAMBEL OAK, Shinnery

BEECH FAMILY — Fagaceae

## GAMBEL OAK — *Quercus gambelii* Nutt.

DESCRIPTION — One of the two oaks in Arizona which sheds its leaves each fall, Gambel oak is a shrub, often forming small dense thickets, or a tree, 6½ to sometimes 50 feet high. The bright green alternate leaves are 2 to 6 inches long, deeply lobed, and usually cut more than half way to the midvein. The pointed, eggshaped acorn is about ½ inch long, and matures the first year. The cup is hairless on the inside, and covers ⅓ to ½ of the acorn.

DISTRIBUTION — Gambel oak is common throughout Arizona, except in the extreme western parts. It occurs in open places on mountain slopes, plateaus, canyons, in yellow pine, pinyon-juniper, and upper chaparral-oak woodland; 5,000 to 8,000 feet elevation. It is a good habitat plant for deer and turkey.

POISONOUS PROPERTIES — Livestock and deer eat the large, abundant leaves whenever available, and the acorns in the fall. It is fairly palatable forage, but, like all oaks, contains tannic acid. Gambel oak is the principal offender in oak poisoning in Arizona, especially among cattle. Poisoning usually occurs in the spring when other forage is lacking. Stock losses result from a diet of 50% or more oak over a considerable period. Gambel oak poisoning occurs in the Prescott area, and other areas where dense stands are formed.

Fig. 35. Gambel oak *(Quercus gambelii)*. Fruiting branch. *a.* Acorn, enlarged.

# SKELETONWEED ERIOGONUM
BUCKWHEAT FAMILY — Polygonaceae

## SKELETONWEED ERIOGONUM — *Eriogonum deflexum* Torr.

DESCRIPTION — Skeletonweed eriogonum is a stiff, flat-topped, bushy annual, ½ to 1½ feet high, reproducing only by seed. The stems are branched above and leafless, except at the base where a few soft, white hairy, heartshaped leaves appear, then wither and drop off. The many small white or pink flower clusters hang downward on short stalks about ⅛ inch long. The 3-angled dark brown seeds (achenes) are ½ inch long or less.

DISTRIBUTION — An abundant and conspicuous weed on dry, disturbed or rocky soil, on roadsides, vacant lots, city streets, waste places, or eroded areas on desert and desert grassland ranges; 150 to 4,000 feet elevation; flowering most of the year, but principally from March to November. An extremely drought-resistant plant, flourishing with little or no water. The stems turn a maroon color in the winter, and are very conspicuous.

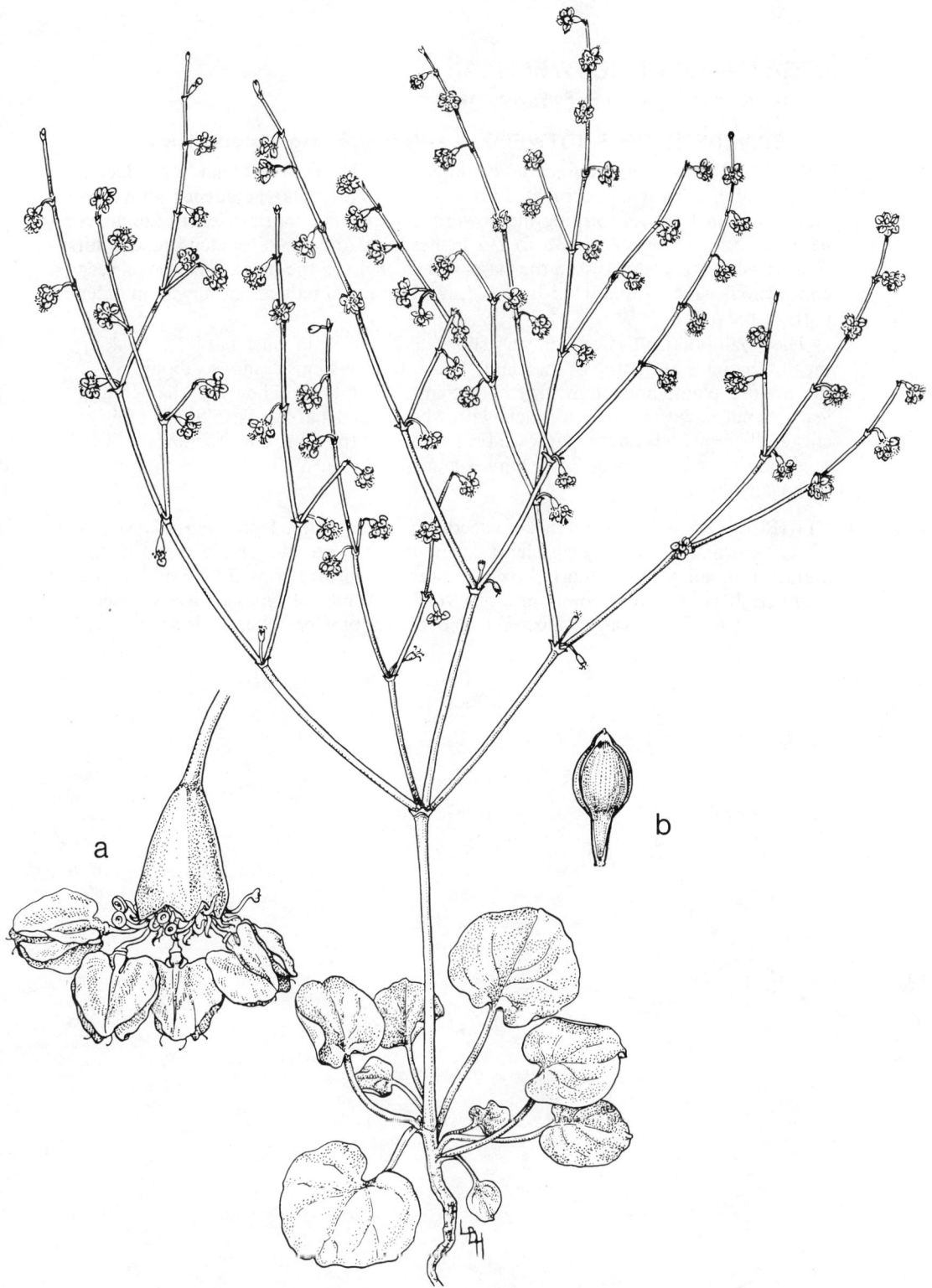

Fig. 36. Skeletonweed eriogonum *(Eriogonum deflexum)*. Flowering plant with basal rosette. *a.* Flower cluster. *b.* Achene with seed enclosed.

# SILVERSHEATH KNOTWEED

BUCKWHEAT FAMILY — Polygonaceae

## SILVERSHEATH KNOTWEED — *Polygonum argyrocoleon* Steud.

DESCRIPTION — A bright green erect annual, 8 inches to 3 feet high, reproducing by seeds alone. The wiry corrugated principal stems are enlarged at each joint, and many branched above, forming the flowering stems. The hairless leaves are alternate, lanceshaped or oblong, ½ to 2½ inches long, and ⅛ to ⅓ inch broad, with thin dryish silvery sheaths at the base which encircle the stem. These are very conspicuous, shining and whitish at first, but become shredded and tawny in older parts of the plant.

The small pinkish flowers, on very short stalks, occur in clusters of 2 to 5 along the flower stems at the top of the plant. The flowering stems compose about half of the mature plant, and often are more than a foot long. They are slender and leafless, but have small green bracts with white sheaths at the base of each flower cluster. The pinkish flower parts (calyx) surround the fruit until the achene with the enclosed seed is shed. The shiny 3-angled achene, about 1/16 inch long, is mahogany colored.

DISTRIBUTION — Silversheath knotweed, an introduction from central Asia, is a weed in alfalfa fields, citrus, lettuce, and other winter vegetable crops in the agricultural valleys of southern Arizona, especially in Maricopa, Pinal, and Yuma counties. It is also very common along roadsides and all types of waste places; 100 to 3,500 feet elevation; flowering November or December to June or July.

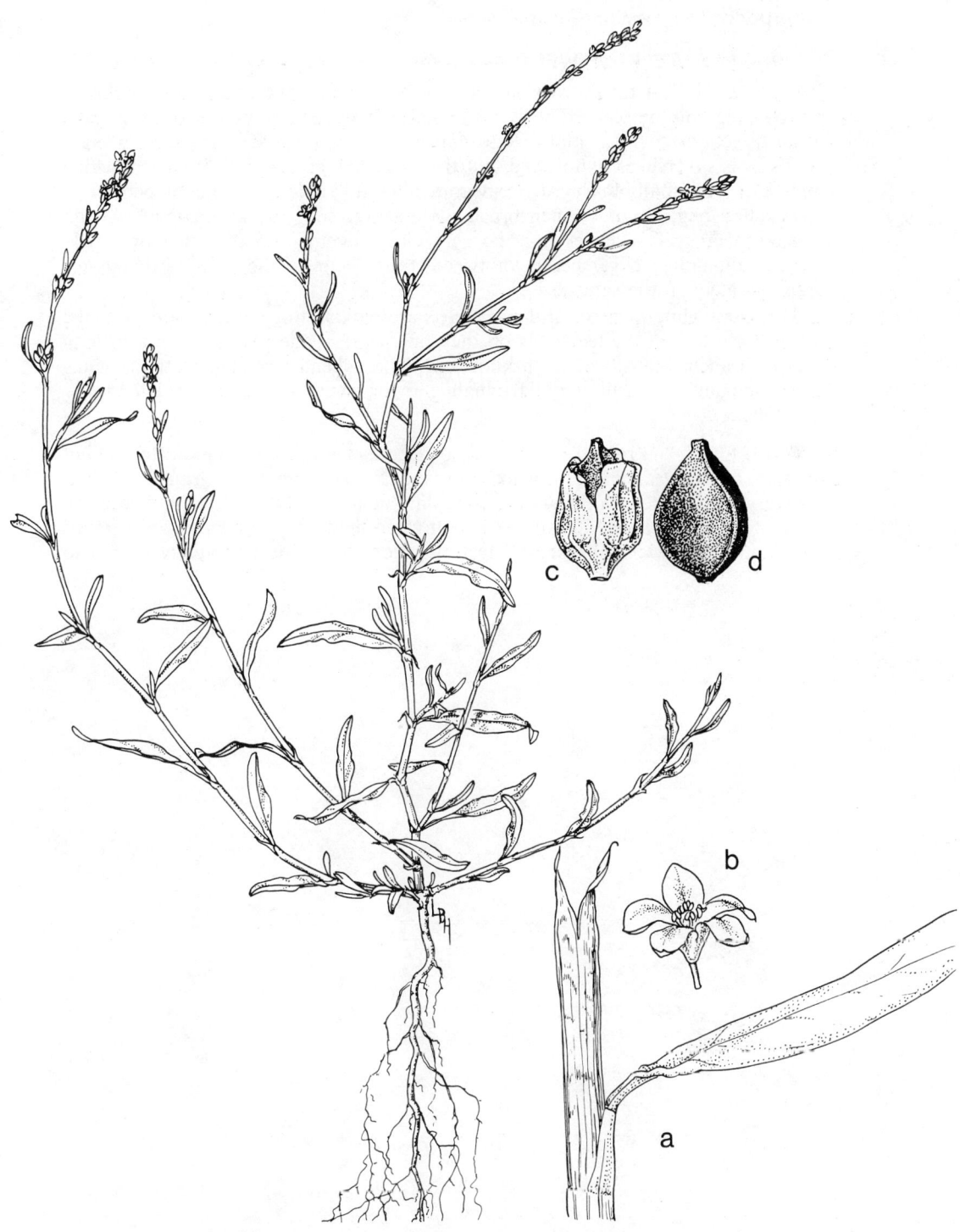

Fig. 37. Silversheath knotweed *(Polygonum argyrocoleon)*. Flowering plant.
*a.* Leaf showing basal sheath. *b.* Flower. *c.* Fruiting calyx. *d.* Achene.

# PROSTRATE KNOTWEED

BUCKWHEAT FAMILY — Polygonaceae

## PROSTRATE KNOTWEED — *Polygonum aviculare* L.

DESCRIPTION — A tough, wiry stemmed, prostrate annual that appears perennial, reproducing only by seeds. The stems branch greatly, spreading out from the root in all directions to form mats ⅓ to 4 feet in diameter. They are seldom erect plants as is silversheath knotweed, but the ends may ascend 3 to 9 inches in rich moist soil. The small bluish green leaves are alternate, oblong or lanceshaped, ¼ to 1½ inches long, about ⅜ inch broad, with a very short stalk, encircled by thin papery sheaths. The leaves may be very close together on the stem or widely spaced, but unlike silversheath knotweed, they occur to the ends of the stems, approximately all the same size.

The small white or greenish flower parts are pinkish tinged. They are arranged in small clusters in the leaf axils on the upper part of the plant, and not in long leafless branches as silversheath knotweed. The achene is 3-angled; the surface, except the angles, is dull, very dark mahogany or blackish, and larger, 1/12 to ⅛ inch long.

DISTRIBUTION — Prostrate knotweed, also Eurasian, is very common throughout the state on dry packed soil of walks, yards, doorways, and waste ground where it is a very hardy weed, withstanding trampling and drought. It is also common in more moist soil in lawns, flower beds, cultivated fields, and eroded or overgrazed mountain meadows; 100 to 8,500 feet elevation; flowering principally March to October.

84

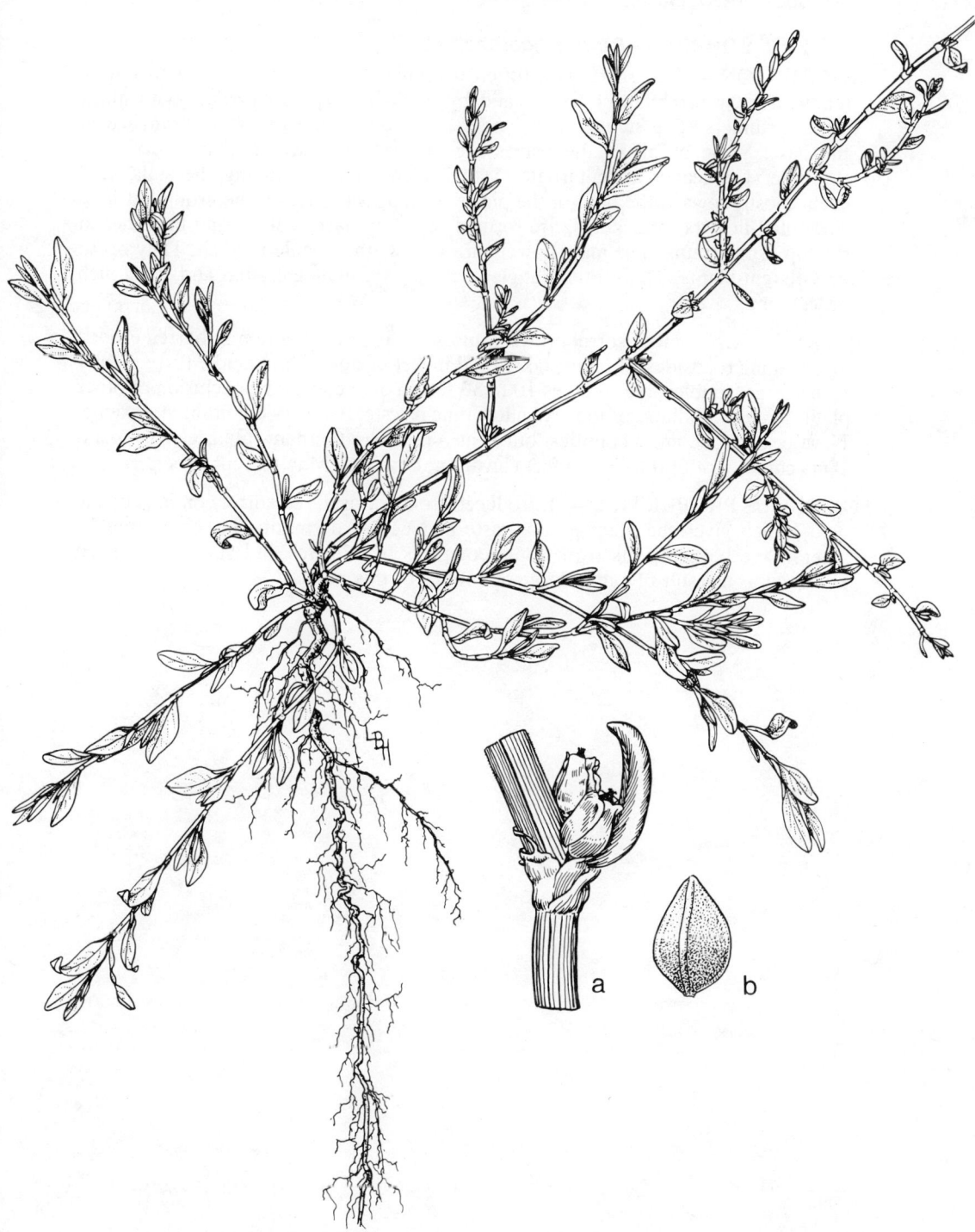

Fig. 38.  Prostrate knotweed *(Polygonum aviculare)*. Plant habit.
*a.* Flower cluster in axil of leaf.  *b.* Achene with enclosed seed.

85

# RED SORREL, sheep sorrel
BUCKWHEAT FAMILY — Polygonaceae

## RED SORREL — *Rumex acetosella* L.

DESCRIPTION — Red sorrel is a tufted perennial ½ to 2 feet high, with few to many stems, reproducing by seed, and by reddish creeping rhizomes with many slender runners. The stems, also slender and reddish, are mostly unbranched at the base. Some or all of the leaves are arrowshaped with a pair of lobes at the base. The leaves are alternate, 1 to 3 inches long including the stalk, and occur mostly toward the base of the plant, forming a rosette on the ground at first. The small flowers, of 2 kinds, are formed on many narrow flowering branches at the top of the stems. The male flowers are yellow, the female reddish. They occur on different plants. The shiny, 3-angled achenes are mahogany red and 1/16 inch or less long.

DISTRIBUTION — Red sorrel grows in moist disturbed soil, often in barren eroded areas, along roadsides, streams, lakes, and in wet meadows and fields. It frequently forms large troublesome colonies 10 to 30 feet in diameter to the exclusion of other plants in oak woodland or mostly yellow pine forests. It is very common in Apache, Navajo, and Coconino counties, but is also found in Graham, Gila, Cochise, and Pima counties; 4,500 to 8,700 feet elevation; flowering May to September.

POISONOUS PROPERTIES — Introduced from Eurasia, red sorrel contains oxalic acid, which gives the plant a sour taste and causes dermatitis in some animals. There have been reports from some countries that horses and sheep have been poisoned as a result of eating large quantities of this plant.

Fig. 39. Red sorrel (*Rumex acetosella*). *a.* Male plant with male flowers. Note the underground rhizomes. *b.* Female branch in fruit. *c.* Achene with enclosed seed.

# CLUSTER DOCK, green dock
BUCKWHEAT FAMILY — Polygonaceae

## CLUSTER DOCK — *Rumex conglomeratus* Murr.

DESCRIPTION — Slender hairless perennial, reproducing by seed and spreading by vertical rhizomes, which often give rise to several stout spindleshaped taproots. The slender stems are 2 to 5 feet high, simple or branched. The principal leaves, oblong to lanceshaped, are 2 to 8 inches long, on stalks ½ to 3 inches long; the upper leaves are much shorter than the lower, and the leaf margins are slightly wavy. The small inconspicuous greenish flowers grow in clusters around the stem from the leaf axils, at intervals of ¼ to 1 inch. The many-branched flowering part is ⅓ to 1½ feet high, and often constitutes as much as half of the plant in age. The flower, similar to all docks, has 6 floral parts; the outer 3 are small, but the inner 3 enlarge as the fruit and seed mature and surround them. In cluster dock, these inner segments are fiddleshaped, about ⅛ inch long, and each bears a grainlike body on its back. The flower stalks are short and jointed near the base. The achene is dark brown and shining, less than ⅛ inch long; the seed is permanently enclosed within.

DISTRIBUTION — Cluster dock, a native of Europe, is a moisture-loving weed occasionally found along mountain streams at mid-elevations and common in summer crops, particularly after irrigation begins. Found in sorghum, cotton, alfalfa, and the vegetable crops, such as lettuce and melons, also common along ditchbanks and low places where extra water collects. In southern Arizona in Maricopa, Pinal, Graham, and Pima counties, and doubtless other counties, since cluster dock has spread as the agricultural lands in Arizona have increased; 1,000 to about 6,500 feet elevation; flowering March to September or October.

Fig. 40.  Cluster dock *(Rumex conglomeratus)*.
*a.* Flower.  *b.* Fruiting calyx. *c.* Achene.

# CURLY DOCK

BUCKWHEAT FAMILY — Polygonaceae

## CURLY DOCK — *Rumex crispus* L.

DESCRIPTION — A restricted noxious weed in Arizona, curly dock is a perennial, 1½ to 4 feet high, with a large deep taproot, reproducing by seed. The smooth fleshy stems die back each fall, and new ones arise each spring. The bluish green lanceshaped leaves are alternate above, and often a large basal rosette is produced late in the fall. They are 3 to 12 inches or more long including the stalk, with noticeably wavy and curly edges.

The small flowers, yellowish green at first, but becoming rosy, then reddish brown, in fruit are borne in a large loose, branching flowering cluster, ½ to 2 feet long, at the top of the stems. These flowers aren't nearly as densely compact as in canaigre. As in all dock, the 6 flower parts do not look like petals. The 3 inner parts become greatly enlarged and heartshaped; they surround the tiny fruit, and have the appearance of wings. At maturity these wings are ⅛ to ¼ inch long, and each usually has a little wartlike thickening on the back. Each flower produces one glossy, triangular achene, about 1/12 inch long.

DISTRIBUTION — Curly dock, an Eurasian introduction, grows in deep moist soil. It is very common along roadsides and waste places, but is a pernicious weed in permanent pastures, irrigation ditches, and in many cultivated crops. It is found throughout the state; from 100 to 8,000 feet elevation, flowering May to October.

Fig. 41. Curly dock *(Rumex crispus)*. Basal portion of plant with
leaves and large taproot. Also upper part of fruiting branch.
*a*. Fruiting calyx. *b*. Achene.

91

# CANAIGRE, wild rhubarb

BUCKWHEAT FAMILY — Polygonaceae

## CANAIGRE — *Rumex hymenosepalus* Torr.

DESCRIPTION — Coarse perennials with stout fleshy stems, 1 to 3 feet high, reproducing by seed. In canaigre, the stems arise from a cluster of 2 to 12 long dahlialike tubers, buried deep in the ground so they are seldom seen. The bright green leaves are hairless, and have thin papery sheaths at the base of the stalk which completely surround the stem (as in all dock species).

The rosette and lower leaves are ⅔ to 2 feet long, and 2 to 4 inches broad, with smooth edges, not wavy as in curly dock. The crowded flower branches form a very dense leafless flower cluster, about ½ to 1 foot long, at the top of the stem. These may superficially resemble those of milo maize. The 6 greenish flower parts are similar to those of curly dock; the 3 inner parts become enlarged and enclose the fruit. These flower parts are ½ to ⅔ inch long at maturity, and do not have a wartlike thickening on the back. The achenes are very similar to those of curly dock, but larger.

DISTRIBUTION — Canaigre is a native plant. Like curly dock, it grows in deep moist, often sandy soil along roadsides, fields, streambeds, and stagnant pools throughout most of Arizona, but isn't nearly as common. It is found at 1,000 to 6,000 feet elevation; flowering very early, in March and April in southern Arizona, and until June northward. Canaigre contains a high percentage of tannin; the Agricultural Research Service, U.S.D.A. has developed improved varieties that offer promise as a crop. The tannin is excellent for leather making.

Fig. 42. Canaigre *(Rumex hymenosepalus).* Basal part of plant
showing leaves and tubers. Also part of a fruiting branch. *a.* Fruiting calyx.

93

# WHEELSCALE SALTBUSH

GOOSEFOOT FAMILY — Chenopodiaceae

## WHEELSCALE SALTBUSH — *Atriplex elegans* (Moq.) D. Dietr.

DESCRIPTION — A bushy much branched annual, densely covered by small silverish mealy scales, and reproducing only by seeds. The erect stems, branching from the base, are 1 to 2 (or 3) feet high, and slightly woody at the base. The thin narrow alternate leaves, ⅜ to 1 or 2 inches long, and 1/12 to ¼ inch broad, often have several small teeth along the edge. Frequently the lower leaf surface is silvery, and the upper greenish.

The small inconspicuous flowers are abundant, and of 2 kinds: male and female. They occur together in small clusters at the base of the leaves, and also form very short spikes at the tip of each stem branch. The seed is permanently enclosed between 2 wheelshaped bracts. The grayish bracts are about ⅛ inch in diameter, with the greenish margin divided into 9 to 20 narrow distinct teeth. A form at Tucson also has a few warty projections on the faces of the bracts.

DISTRIBUTION — Wheelscale saltbush, a native, is abundant in hard-packed, heavy alkaline, dry caliche, or moist fertile soil in cultivated fields, citrus groves, roads, city streets, ditches, and waste places in southern Arizona; 100 to 3,500 feet elevation; flowering May to November, but mostly in late summer.

## SALTON SALTBUSH — *Atriplex fasciculata* Wats.

DESCRIPTION — Closely resembles wheelscale saltbush, but the stems which are branched from the base are prostrate with the tips ascending, rather than erect. This gives the growing plant an entirely different appearance. The leaves are more spatulashaped, broad and rounded above, narrowed at the base, often ½ inch broad, with fewer or no teeth along the margins. The bracts enclosing the seed are also wheelshaped; the greenish margins are divided not into distinct countable teeth, but into minute teeth too short and indistinct to count even with the aid of a lens.

DISTRIBUTION — Salton saltbush grows in the same type of places, has the same distribution and flowering period as wheelscale saltbush, but isn't nearly as common.

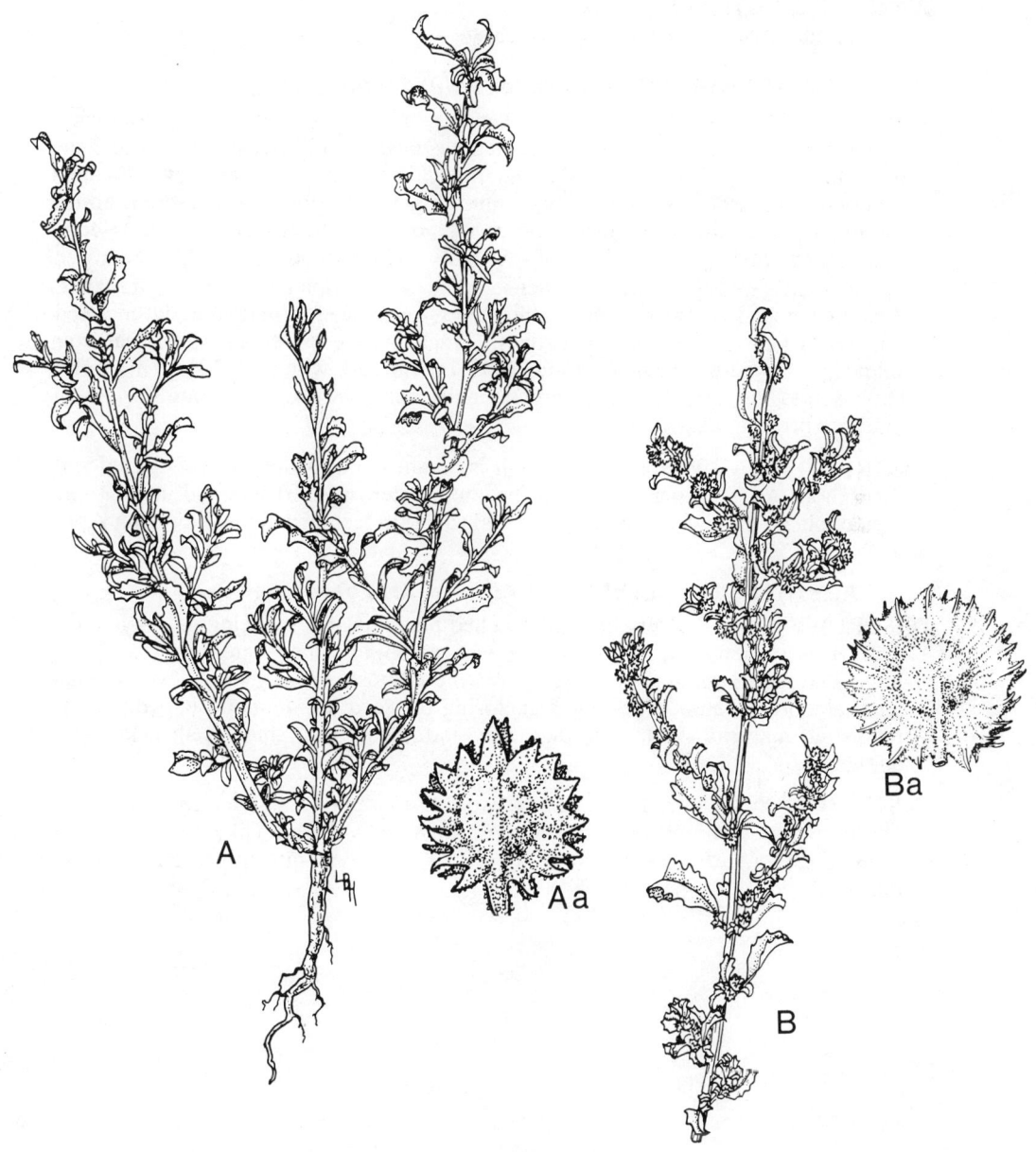

Fig. 43. *A.* Wheelscale saltbush *(Atriplex elegans)*. Leafy plant.
*Aa.* Fruiting bracts, enclosing a seed. *B.* Salton saltbush *(A. fasciculata)*.
Fruiting branch. *Ba.* Fruiting bracts.

# WRIGHT SALTBUSH

GOOSEFOOT FAMILY — Chenopodiaceae

## WRIGHT SALTBUSH — *Atriplex wrightii* Wats.

DESCRIPTION — A tall robust annual with somewhat woody roots, reproducing only by seeds. The stems are erect, sometimes bushy, and branched from the base or branching above only, 2 to 4 feet high. The mature leaves, conspicuously silvery beneath from meallike particles, and bright green above, are broadest above the middle, with the tip blunt or somewhat pointed. They are alternate, ¾ to 3 inches long, and ¼ to 1 inch broad, with the edges smooth, wavy, or toothed.

The male flowers are in short dense spikes on conspicuous leafless branches, 2 to 12 inches long, at the top of the stems. Each seed is permanently enclosed between a pair of greenish or yellowish bracts. These bracts are somewhat fan-shaped, or broadly triangular, about 1/12 inch broad, with 3 to 7 short irregular teeth across the top. The faces are strongly ribbed, with or without very short knoblike projections.

DISTRIBUTION — This native saltbush is common in alkaline and disturbed soil throughout Arizona, particularly in the southeastern part. It is found in the same type of places as wheelscale saltbush; 100 to 7,000 feet elevations; flowering April to October.

## AUSTRALIAN SALTBUSH — *Atriplex semibaccata* R. Br.

DESCRIPTION — A silvery, much branched perennial, reproducing only by seeds. The stems are somewhat woody at the base, reclining or prostrate, and 1 to 3 feet long. The leaves, similar to those of Wright saltbush, usually have several small teeth along each edge. The bracts enclosing the seed are diamondshaped, fleshy, reddish at maturity, ⅛ inch in diameter, and have 2 to 5 small teeth near the 2 corners.

DISTRIBUTION — Introduced from Australia, this plant has escaped cultivation, and become a common weed in southern Arizona. Valuable (like all saltbushes) as livestock forage, and also as a soil binder, it has spread into cultivated fields and become a pest in some areas.

Fig. 44. *A.* Wright saltbush *(Atriplex wrightii).* Flowering branch.
*Aa.* Male flowers, enlarged. *Ab.* Cluster of three fruits. *B.* Australian
saltbush *(A. semibaccata).* Fruiting branch. *Ba.* Enlarged fruit.

97

# FIVEHOOK BASSIA, smother weed
GOOSEFOOT FAMILY — Chenopodiaceae

## FIVEHOOK BASSIA — *Bassia hyssopifolia* (Pall.) Kuntze

DESCRIPTION — A tall stout erect annual, 2 to 5 feet high, reproducing only by seeds. The plant is branched, commonly with one tough principal stem from which many stiff branches arise, often starting near the base. The flat narrow pointed leaves are alternate, mostly ¼ to ½ inch long, and 1/12 to ⅛ inch broad, with smooth edges. Each has a cluster of shorter leaves in its axil, or a short spike with many small leaves and flowers, very crowded, and woolly with yellowish hairs. These flower spikes are longer at the tips of the stems, sometimes forming conspicuous branches 2 to 12 inches long at the top. The inconspicuous flowers have no petals.

There is a 5-parted, thin, yellowish hairy fruit, which encloses the seed until it matures. Each of the 5 parts develops a yellowish curved spine, hooked at the tip. The weak spines are 1/16 to 1/12 inch long at maturity, but are obvious in very young flowers. The oval seeds are 1/16 to 1/12 inch long, with a dark gray margin.

DISTRIBUTION — Fivehook bassia is a native of Asia. It grows in alkaline wastelands and disturbed soil of roadsides and ditch banks, but is sometimes found in adjacent cultivated fields. It is locally abundant in Maricopa and Yuma counties, and now has spread northward to Mohave, Coconino, and Navajo counties; 100 to 5,700 feet elevation; flowering July to October. It is a late summer host of the leafhopper genus *Lygus*.

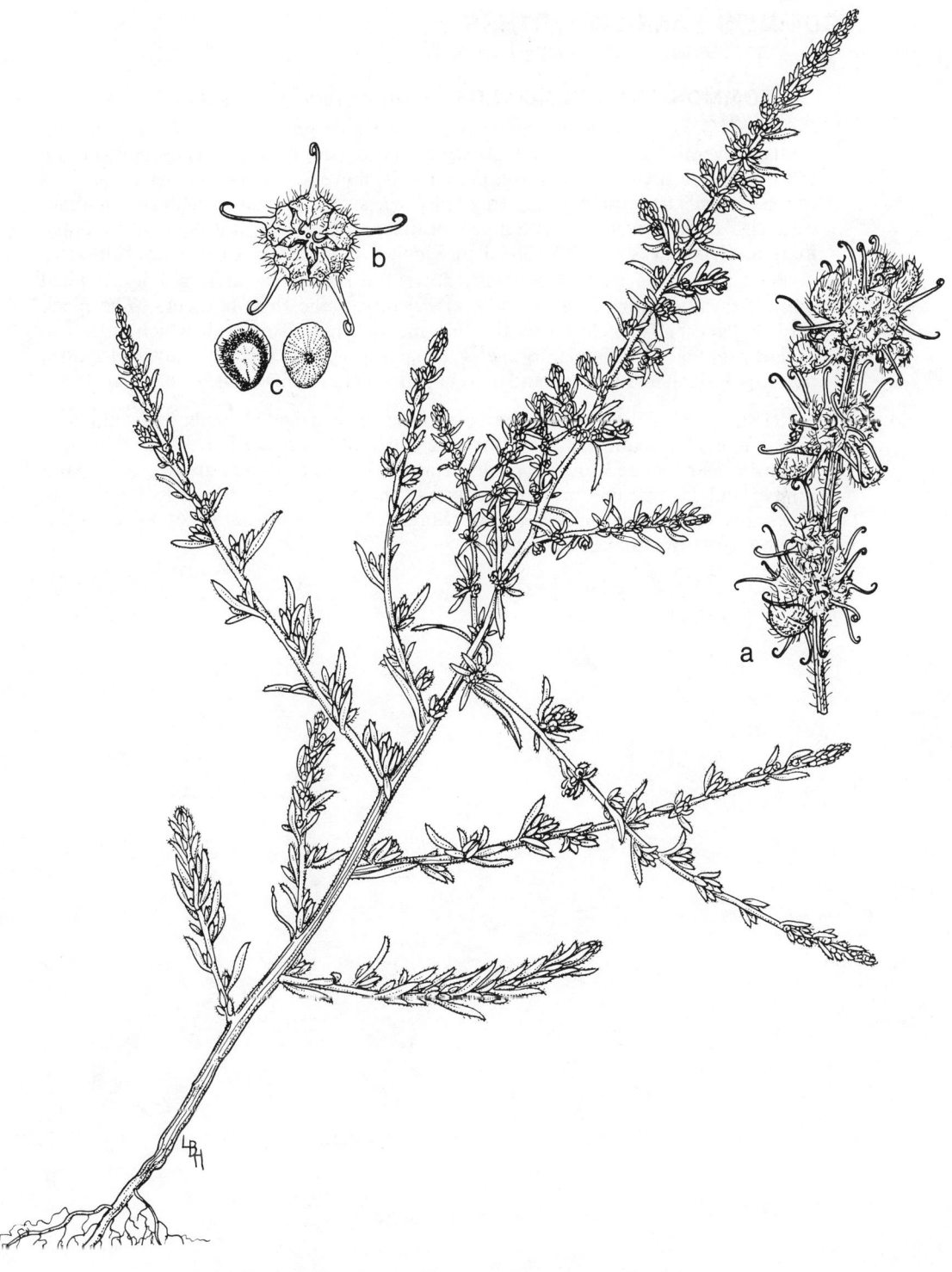

Fig. 45. Fivehook bassia *(Bassia hyssopifolia)*. Plant habit.
*a.* Enlarged portion of fruiting branch. *b.* Fruiting calyx.
*c.* Two views of a seed.

# COMMON LAMBSQUARTERS
GOOSEFOOT FAMILY — Chenopodiaceae

## COMMON LAMBSQUARTERS — *Chenopodium album* L.

DESCRIPTION — Common lambsquarters is a pale green annual with one main stem or several, 1 to 4 or 6 feet high, which reproduces by seeds. The plant is more or less white mealy throughout, particularly the flowers, and usually the lower sides of the leaves. The variable leaves may be lanceshaped and smoothedged, or somewhat egg- or wedgeshaped, with a pair of lobes at the base, and often with toothed margins. The leaves are alternate, 1 to 5 inches long, and ½ to 2 inches broad.

The small inconspicuous greenish flowers are stalkless, arranged in crowded clusters on the short flower clusters at the tips of the stem branches. The black seed is persistently enclosed by the thin membranous fruit wall, which gives it a dull appearance. If this membrane is scraped away, the seed is shiny and glossy, about 1/16 inch in diameter, and diskshaped, with a notch on one side.

DISTRIBUTION — Lambsquarters is a nuisance in irrigated lands and cultivated crops. It is also found on river bottoms and eroded areas of overgrazed ranges, brush burns or logged forest openings, in the desert or desert grassland, pinyon-juniper and yellow pine throughout most of the state; 100 to 9,500 feet elevation; flowering from early summer to fall, May to October. A native of Europe, this weed is good livestock feed.

Fig. 46. Common lambsquarters *(Chenopodium album)*.
Flowering plant. *a.* Enlarged leaf. *b.* Enlarged seed.

# NARROWLEAF GOOSEFOOT

GOOSEFOOT FAMILY — Chenopodiaceae

## NARROWLEAF GOOSEFOOT — *Chenopodium desiccatum* A. Nels. (*C. pratericola* Rydb.)

DESCRIPTION — A stout grayish annual, 1 to 4 feet high, which usually branches well above the base, and reproduces only by seeds. The lanceshaped or oblong pointed leaves are short stalked, densely whitish, mealy on the lower surface and greener above, ½ to 1½ or 2 inches long, 1/12 to ⅓ inch broad, with both large and small leaves occurring together. The leaves may be 1-veined with smooth edges, or 3-veined at the base with 2 indistinct, or short but obvious lobes, 1 on either side toward the base.

The small stalkless flowers, also densely covered by whitish mealy scales, are crowded on short branches in the upper leaf axils, and in long, leafless flowering branches at the top of the plant. The black, shiny diskshaped seed, about 1/25 inch in diameter, drops out of the thin, papery covering when mature.

DISTRIBUTION — This native goosefoot grows in moist or dryish alkaline or disturbed soil. A common weed throughout most of Arizona, and a pest in field crops in the southern part of the state, also found in irrigation ditches, waste places, river flats, and roadsides; 100 to 8,500 feet elevation; flowering May to September.

## SLIMLEAF GOOSEFOOT — *Chenopodium leptophyllum* Nutt.

DESCRIPTION — An annual, closely resembling narrowleaf goosefoot, and often confused with it. Differing in that the leaves are narrow, 1-veined, with smooth margins. The seed is permanently enclosed in the thin colorless covering, and thus the surface is dull, not shiny.

DISTRIBUTION — Slimleaf goosefoot is also a native weed, found principally in northern and central Arizona, along roadsides, washes, waste places, and also in barren areas in pinyon-juniper, and yellow pine ranges in Apache, Navajo, Coconino, Yavapai, and Pinal counties; 2,500 to 8,500 feet elevation; flowering August to September.

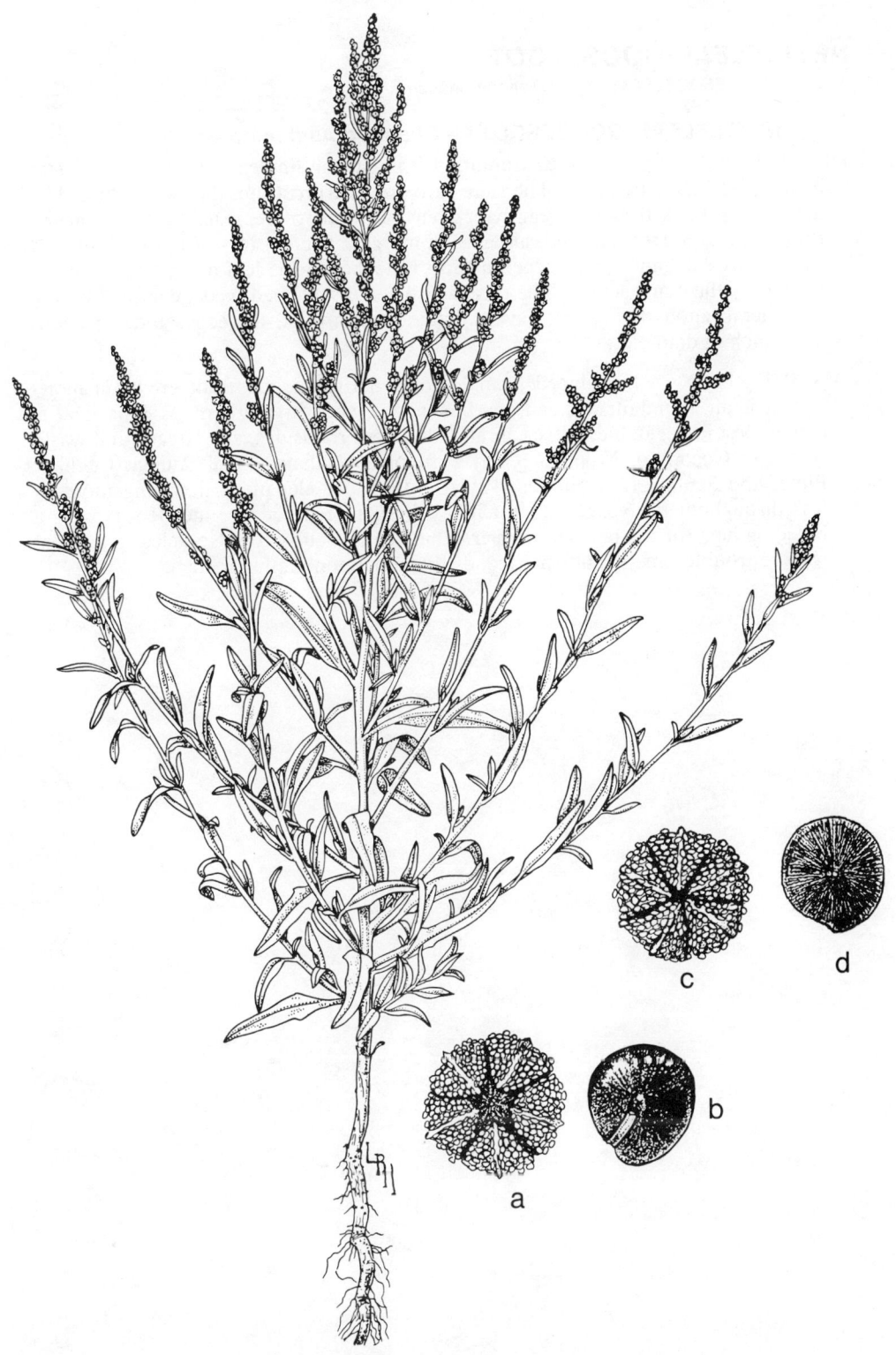

Fig. 47. Narrowleaf goosefoot *(Chenopodium desiccatum)*. Flowering plant. *a.* Fruiting calyx. *b.* Seed. *c.* Slimleaf goosefoot *(C. leptophyllum)*. Fruiting calyx. *d.* Seed.

103

# NETTLELEAF GOOSEFOOT

GOOSEFOOT FAMILY — Chenopodiaceae

## NETTLELEAF GOOSEFOOT — *Chenopodium murale* L.

DESCRIPTION — Coarse bushy annual with a strong unpleasant odor, 1 to 3 feet high, reproducing by seeds. The stems are erect or prostrate, then ascending. The dark green, thick triangular eggshaped leaves are alternate, usually pointed at the tip, with 1 to 8 irregular teeth along each margin. The clusters of flowers, covered by a mealy substance, are borne on short branches in the leaf axils, and often are hidden by the dense leaves. The tiny, dull black diskshaped seeds, enclosed by the thin membranous fruit wall, have a definite rim around the edge, and are about 1/18 inch in diameter.

DISTRIBUTION — Introduced from Europe, nettleleaf goosefoot grows in moist soil. It is an abundant winter to early summer weed in southern Arizona, and is a great pest in vegetables, citrus, alfalfa, and flax, roadsides, city streets, and waste places in Coconino, Yavapai, Gila, Maricopa, Graham, Pinal, Yuma, Cochise, Pima, and Santa Cruz counties; 100 to 8,000 feet elevation; flowering more or less throughout the year, but principally from December to June. It serves as a breeding host for the beet leafhopper, which causes curly top. Nettleleaf goosefoot is more troublesome in cultivated lands than common lambsquarters.

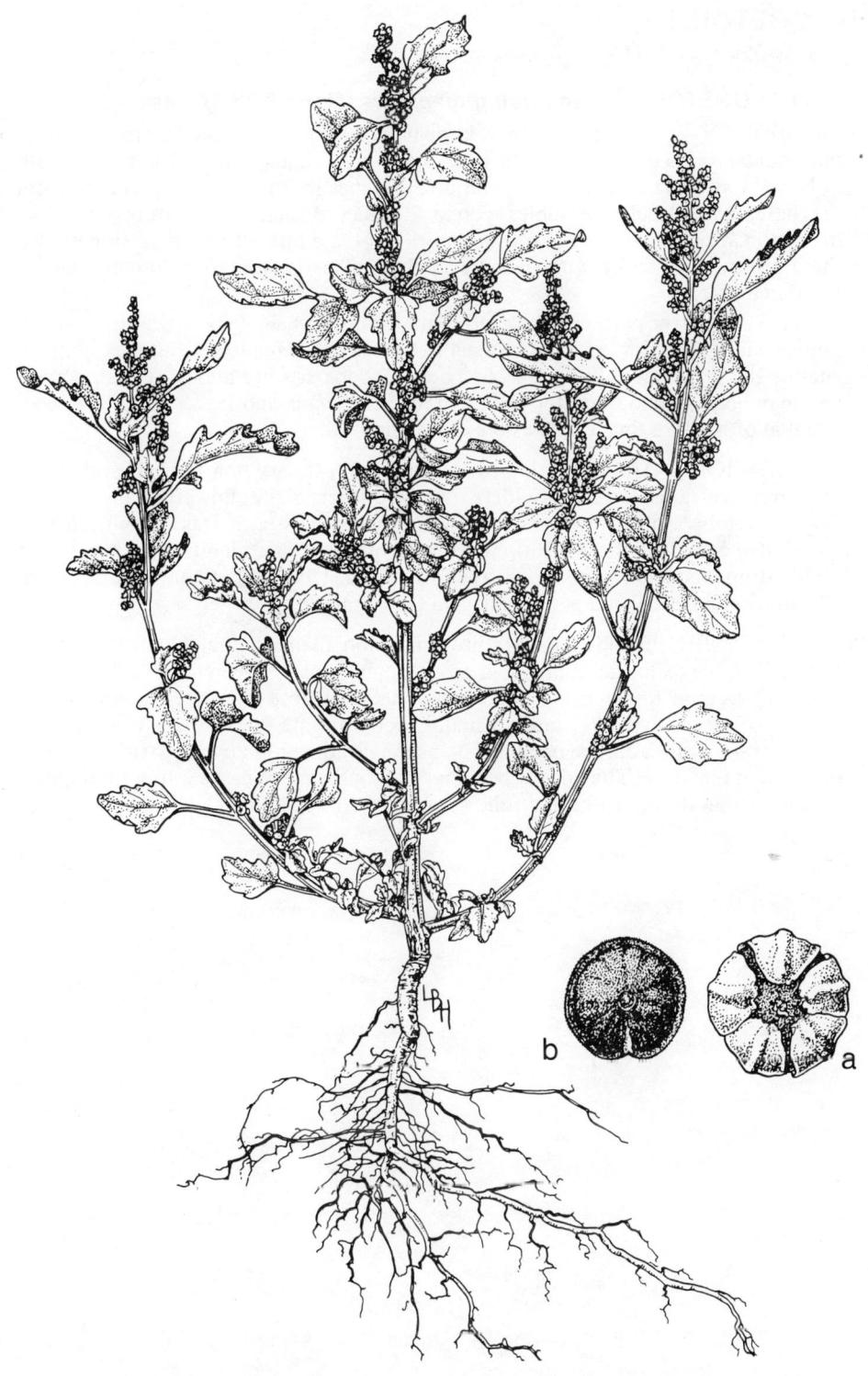

Fig. 48. Nettleleaf goosefoot *(Chenopodium murale)*. Flowering plant.
*a.* Fruiting calyx. *b.* Achene with enclosed seed.

# HALOGETON, barilla

GOOSEFOOT FAMILY — Chenopodiaceae

## HALOGETON — *Halogeton glomeratus* (M. Bieb.) C. A. Mey

DESCRIPTION — Halogeton is not known to occur in Arizona at this writing, but doubtless it is only a matter of time, since it is common along the border both in Nevada and Utah. It is a fleshy annual, 2 inches to 2 feet high, branching from the base, and closely resembles young Russian thistle. The mature leaves of halogeton are soft, and end in a white hairlike bristle instead of a rigid spine. Also there are tufts of kinky whitish hairs in the leaf axils, which are not present in Russian thistle.

The 5 dry flower parts enlarge, and at maturity form showy yellowish to reddish fanlike wings. The flower parts are usually thought of as the fruits, but they actually enclose the tiny fruit with its 1 seed. These flower parts are about ¼ inch across, and may be so abundant in the fall as to hide the stems and leaves. The tiny seed, like that of Russian thistle, shows the coiled embryo.

DISTRIBUTION — Halogeton is a weed of the dry deserts, on barren eroded soil of overgrazed ranges, road shoulders, or any disturbed site; flowering and fruiting July to October with fruits persistent. A native of Russia, it was introduced into the United States in about 1930, and apparently was first identified in Nevada in 1934. It now covers millions of acres in Nevada, Utah, and Idaho, and is found also in Montana, Wyoming, Oregon, and California.

POISONOUS PROPERTIES — Mature halogeton plants contain high concentrations of soluble oxalic acid salts, and are extremely toxic. Sheep are poisoned more often than cattle under usual conditions. A lethal dose for sheep is about 1½ pounds of green plant. Poisoning usually occurs in late fall or early winter, since snow or rain washes out the poison. The poison is not cumulative; a toxic dose must be eaten at one time. The animal becomes dull and cannot move 2 to 4 hours after eating a lethal dose; death may follow within 6 to 10 hours.

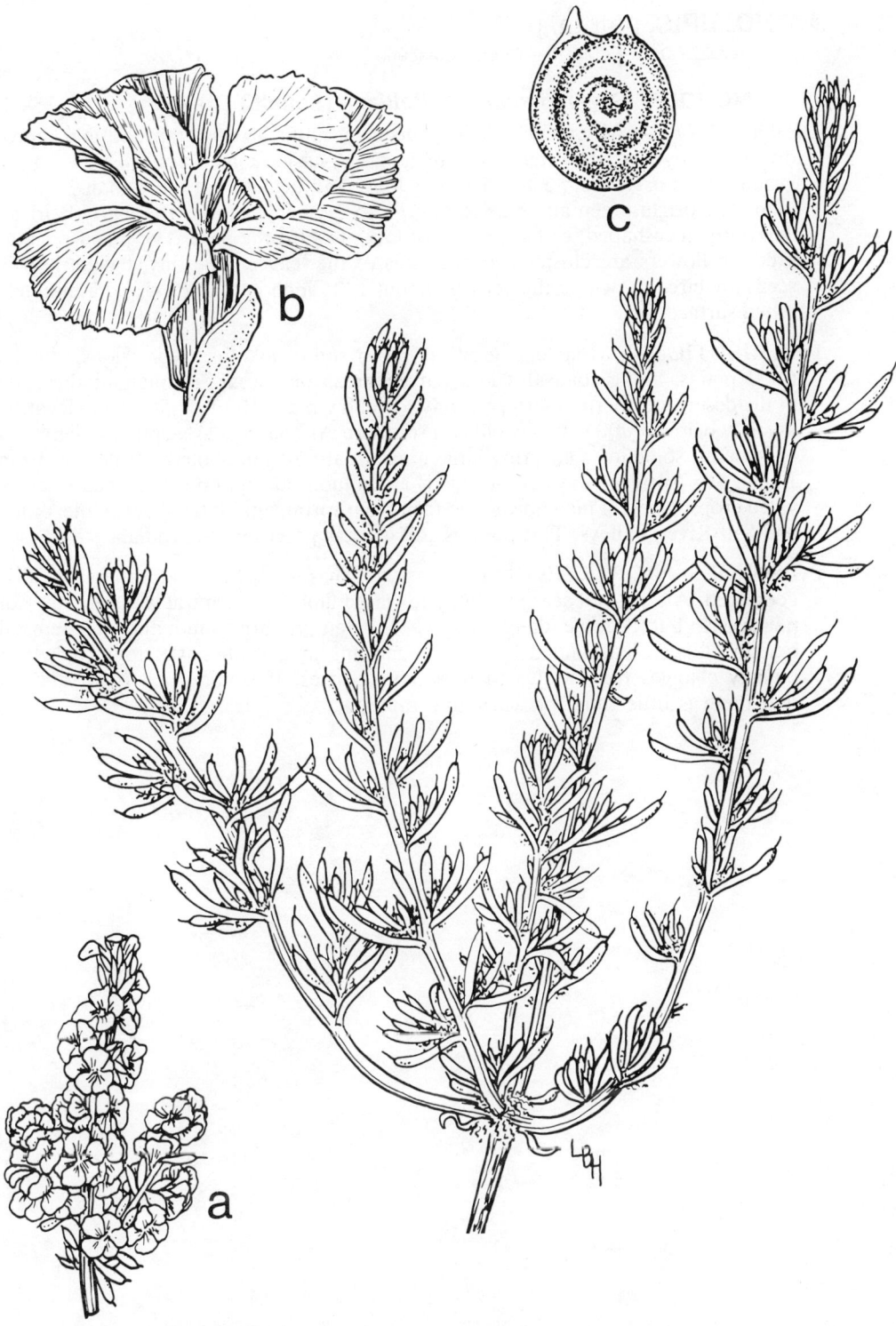

Fig. 49. Halogeton *(Halogeton glomeratus).* Flowering branch.
*a.* Enlarged section of a flowering branch. *b.* Fruit. *c.* Seed.

107

## MONOLEPIS, patota, patata
GOOSEFOOT FAMILY — Chenopodiaceae

### MONOLEPIS — *Monolepis nuttalliana* (Schult.) Greene

DESCRIPTION — Monolepis is a low succulent annual with a taproot, and reproduces only by seed .The stems are erect, spreading, or sometimes prostrate on the ground, then ascending, 3 to 15 inches long, and usually not more than 8 inches high. The bright green alternate leaves, ⅓ to 2½ inches long, are nearly hairless. They are lanceshaped, or have a pair of lobes toward the base. The inconspicuous greenish flowers are clustered in the axils of the leaves. The tiny, flat dark gray seeds are circular, with a thicker rim, about 1/25 inch in diameter, and a minutely pitted surface.

DISTRIBUTION — Monolepis grows in moist soil of alkaline depressions, alluvial flood plains, along roadsides and barren areas on mesas, throughout the state in the desert, northern desert, or rarely to yellow pine; 100 to 7,500 feet elevation (mostly below 3,000). It is prolific in southern Arizona in early spring, where it is ephemeral, sometimes covering large areas in almost pure stands. It flowers from January to April or May. Monolepis is a common but unimportant weed in cultivated crops, where it may flower and fruit from spring until late fall, as in the Yuma and Salt River Valleys. This plant is still used as greens by the Indians.

POISONOUS PROPERTIES — Livestock relish monolepis, and ordinarily it is good feed. After a wet season it contains large amounts of nitrates, which are not poisonous, but become dangerous when eaten in large amounts. A chemical reaction takes place, aided by microorganisms in a ruminant's stomach, which quickly changes the nitrates to poisonous nitrites. Poisoning occurs so quickly that there is little time for warning symptoms.

Fig. 50. Monolepis *(Monolepis nuttalliana)*.
Flowering plant. *a.* Fruit. *b.* Seed.

# RUSSIAN THISTLE, tumbleweed
GOOSEFOOT FAMILY — Chenopodiaceae

## RUSSIAN THISTLE — *Salsola kali* L. var. *tenuifolia* Tausch.

DESCRIPTION — A restricted noxious weed in Arizona, Russian thistle is an intricately branched, bushy globular annual, ½ to 6 feet high, with ridged and often reddish stems, reproducing only by seed. At maturity, the hard, prickly plant breaks at the ground level, becoming a "tumbleweed." The grasslike seedlings and the young plants are fleshy and tender, with alternate, narrow pointed leaves, ½ to 2 inches long. These leaves drop off; the short, stiff mature leaves are awlshaped, and end in a spine.

The tiny whitish flowers are clustered at the base of the leaves along the upper branches. There are no petals, but the 5 dry flower parts enlarge, and each develops a large veiny wing. These meet to form a cover over the topshaped, reddish, slightly winged fruit. Each fruit has 1 gray to brownish yellow seed, with the coiled embryo visible.

DISTRIBUTION — Russian thistle is one of the most prolific and obnoxious weeds throughout the state. Abundant in southern Arizona in irrigated areas, waste grounds, and river bottoms; also common in small grains. In northeastern Arizona, it is very common on overgrazed ranges and pastures in grasslands, chaparral, pinyon-juniper, and frequently in yellow pine; 150 to 7,000 feet elevation; flowering May to October or November.

A native of Russia, this plant was brought into the United States in flax seed about 100 years ago, and has spread very rapidly. It is a prolific seeder: one plant may produce thousands of seeds. The seeds remain viable for years, and are scattered as the plant rolls along. It is a host plant for the sugarbeet leafhopper, which carries the virus causing curly top in beets. It is also the source of "blight" in other crop plants such as tomatoes, spinach, and beans.

POISONOUS PROPERTIES — May store toxic amounts of nitrates after periods of fast growth.

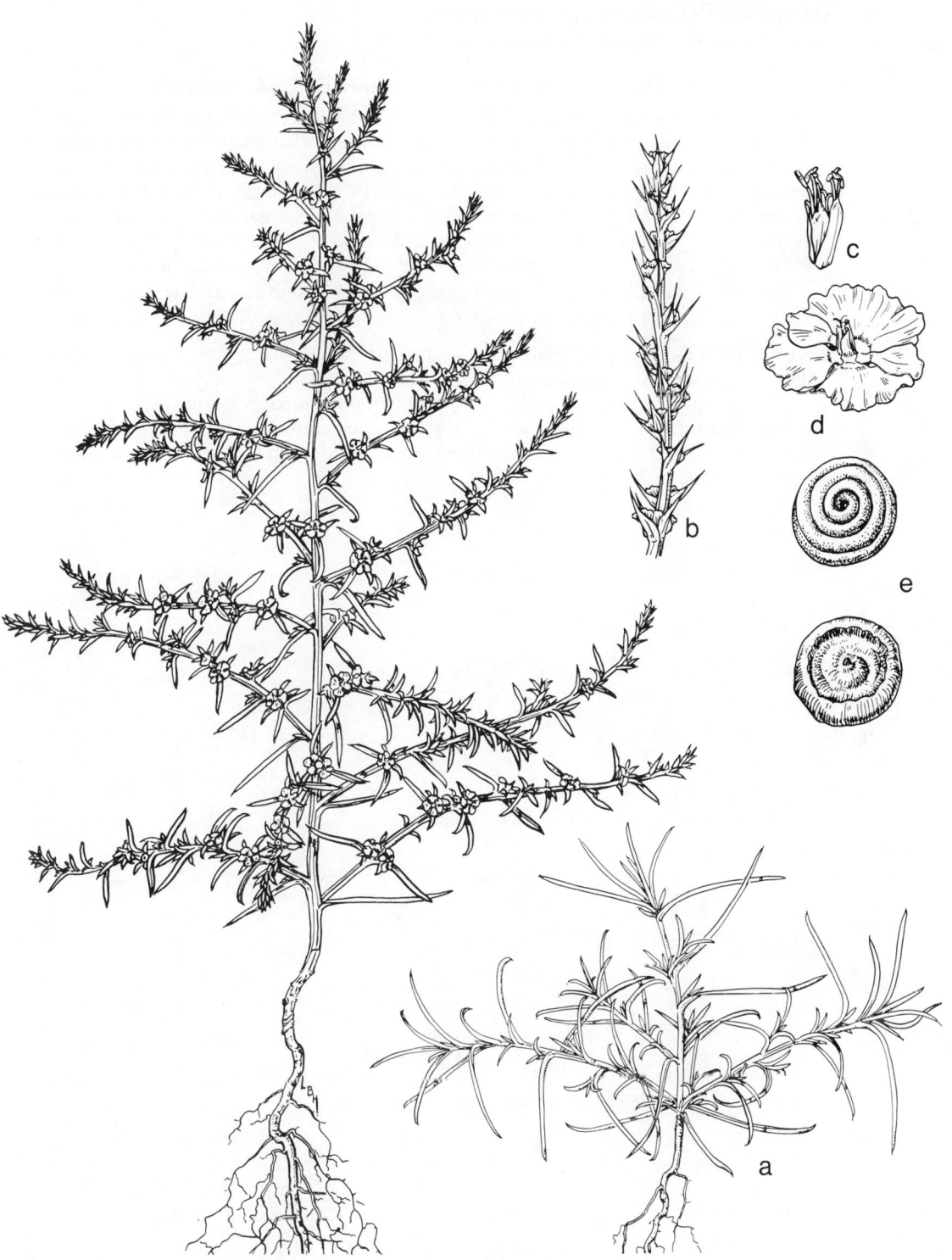

Fig. 51. Russian thistle *(Salsola kali* var. *tenuifolia).* Plant habit.  *a.* Seedling.
*b.* Part of fruiting branch.  *c.* Flower.  *d.* Fruiting calyx.  *e.* Seed.

111

# KHAKIWEED, creeping chaffweed
PIGWEED FAMILY — Amaranthaceae

## KHAKIWEED — *Alternanthera pungens* H. B. K. (*A. repens* [L.] Ktze.)

DESCRIPTION — Khakiweed is a perennial which reproduces by seeds and thick, deep seated, woody vertical roots. The stems are prostrate or drooping on the ground, forming mats ½ to 2 feet in diameter. The plant is covered with jointed distinct hairs, especially at the stem nodes and on the backs of the stiff papery flower parts. The latter hairs are minutely barbed. The glossy leaves are opposite, ½ to 1½ inches long, appearing hairless, but usually with scattered hairs, particularly on the stalks. The 2 leaves of the same pair are often very unequal in size.

The small whitish flowers are in dense clusters around the stem at the base of the leaves. The rounded seeds are light reddish brown and shining.

DISTRIBUTION — Khakiweed, a native of tropical America, has become one of the most pernicious pests in southern Arizona lawns in the last few years. It is very aggressive, and often covers large areas in Bermudagrass lawns, crowding out the Bermudagrass. Found also along roadsides, city streets, gardens, and cultivated fields in Cochise, Pima, and Santa Cruz counties; up to 5,500 feet elevation; flowering June to November.

112

a

b

Fig. 52. Khakiweed *(Alternanthera pungens)*. Flowering plant showing prostrate habit. *a.* Flower cluster. *b.* Seed.

# TUMBLE PIGWEED, tumbleweed
PIGWEED FAMILY — Amaranthaceae

### TUMBLE PIGWEED — *Amaranthus albus* L.

DESCRIPTION — A bushy branched annual ½ to 4 feet high, with light green or whitish stems, which reproduces by seeds. The bright green alternate leaves, 1 to 3 inches long and oblong or spatulashaped, are often reddish purple beneath, with the veins and margins white edged, sometimes with a short bristle at the tip.

The short greenish flowers are in short narrow clusters in the leaf axils; the 3 awlshaped spiny bracts below each flower are much longer. The little papery fruit opens by a circular line, and the top comes off like a lid. The shiny diskshaped seeds are dark reddish brown or black, about 1/25 inch in diameter, with a minutely roughened surface. At maturity the large globeshaped plants are stiff, bristly, and sticky to the touch. They often are broken off at the ground level, and carried about by the wind as a tumbleweed.

DISTRIBUTION — Tumble pigweed is a common weed throughout Arizona, and a pest in cotton, flax, and other cultivated fields. It also grows in river bottoms, roadsides, waste places, and eroded or rocky slopes on rangelands; 100 to 8,000 feet elevation; flowering May to November. This weed was introduced from tropical America.

### PROSTRATE PIGWEED — *Amaranthus graecizans* L.
### (*A. blitoides* Wats.)

DESCRIPTION — Prostrate pigweed is very similar to tumbling pigweed. The stems, however, instead of being erect, are prostrate, forming mats on the soil ½ to 2 feet long, and are often pink or purplish rather than pale green. The 3 spiny bracts at the base of each flower are only slightly longer than the flowers, and the seeds are shining black.

DISTRIBUTION — Prostrate pigweed occurs throughout the state in cultivated fields. It is also found along roadsides, river bottoms, mesas, washes, alkaline sinks, railroad tracks, and denuded areas in overgrazed ranges in mesquite, oak, or pine forests; 100 to 8,200 feet elevation; flowering May to November.

Fig. 53. Tumble pigweed *(Amaranthus albus).* Plant habit. *a.* Stem with flower clusters. *b.* Seed emerging from papery fruit which opens by a circular line.

# FRINGED PIGWEED

PIGWEED FAMILY — Amaranthaceae

## FRINGED PIGWEED — *Amaranthus fimbriatus* (Torr.) Benth.

DESCRIPTION — An erect annual, sometimes bushy with many branches, 1 to 2 (or 3) feet high. The leaves are alternate, lanceshaped or somewhat eggshaped, and pointed at the top. They have smooth edges and a short stalk, 1 to 2 inches long.

The small flowers are of 2 kinds, male and female, with both kinds on the same plant. The male flowers are few, and occur together in clusters in the leaf axils, at the stem tips, and at the top of the plant in long nearly leafless spikes. The many female flowers are very conspicuous and pretty at maturity. The 5 flower parts enlarge, and each spreads out above into a thin pinkish white papery fanshaped structure, the edges of which are slightly fringed. This structure encloses the little thin-walled fruit, whose top falls away as a lid, shedding the tiny seed when it is mature. The oval seed is shiny, reddish black, and about 1/25 inch in diameter.

DISTRIBUTION — Fringed pigweed is a southwestern native plant, growing in dry sandy or rocky soil of desert washes, mesas, and roadsides. It has spread into the cultivated fields, particularly in newer agricultural areas, as in Avra and Yuma valleys; common throughout southern Arizona; northward to Mohave and western Coconino counties; 100 to 4,000 feet elevation; flowering July to October.

Fig. 54. Fringed pigweed *(Amaranthus fimbriatus)*. Flowering plant and branch from more robust plant showing leaf variation. *a.* Mature female flower. *b.* Seed.

# PALMER AMARANTH, carelessweed

PIGWEED FAMILY — Amaranthaceae

## PALMER AMARANTH — *Amaranthus palmeri* Wats.

DESCRIPTION — Palmer amaranth is a tall stout summer annual 1 to 6 (exceptionally 15) feet high, reproducing by seeds. There is 1 thick principal stem, and the lateral branches are usually short. The stem often turns red, particularly in age. The lanceolate or eggshaped, sometimes variegated leaves are alternate, hairless, 2 to 8 inches long including the long stalks, ½ to 2½ inches broad, with very prominent whitish veins on the lower surface.

The inconspicuous flowers are of 2 kinds, male and female, borne on separate plants in long leafless branching spikes at the top of the plant. The slender central flowering tassel, ½ to 1½ feet long, is much longer than any of the side branches at its base. Both male and female flower parts become stiff and spiny at maturity. The dark reddish brown seed is oval, and about 1/16 inch long.

DISTRIBUTION — One of the most common weeds in moist disturbed soils in central and southern Arizona, it also occurs in the northern part of the state. It is a great pest in and around cultivated fields, ditch banks, roadsides, and irrigated pastures, and in all summer crops, especially serious in cotton and sorghum. It also abounds in river bottoms, canyon beds, and on overgrazed ranges in the desert and desert grassland; 100 to 5,500 feet elevation; flowering all summer, May to November.

The pollen of palmer amaranth often causes hay fever. The plant is relished by livestock in all stages of growth, and is sometimes cut for hay or put into silos. Many stockmen consider it good feed.

POISONOUS PROPERTIES — Palmer amaranth contains nitrate varying from a trace to over 9 percent. As in monolepis, the nitrate is not poisonous, but can be changed quickly into the toxic nitrite by enzymatic action.

118

Fig. 55.  Palmer amaranth (*Amaranthus palmeri*). Flowering plant and flowering branch. *a*. Basal leaf. *b*. Male flower showing stamens. *c*. Female flower. *d*. Seed.

# SMALL MATWEED

PIGWEED FAMILY — Amaranthaceae

## SMALL MATWEED — *Brayulinea densa* (Humb. & Bonpl.) Small

DESCRIPTION — A native plant closely resembling khakiweed, but the matted stems are always very close to the ground. The leaves are smaller, ¼ to ½ inch long; the stems, under leaf surfaces, and flower clusters are white woolly. The long, unjointed, weak hairs are not distinct, but soft and entangled.

DISTRIBUTION — Not a city weed like khakiweed, small matweed thrives on dry sandy, hard clay, or rocky soil in hot sun or shade on dirt roads, trails, and especially on overgrazed ranges and waste places around ranches; in desert grassland, pinyon-juniper, and oak-pine association; in southeastern Arizona, Cochise, Greenlee, Graham, Gila, Pima, and Santa Cruz counties; 3,800 to 6,800 feet elevation; flowering June to October. This perennial weed is becoming a serious pest on some ranges in Cochise and Santa Cruz counties.

Fig. 56. Small matweed *(Brayulinea densa)*. Flowering plant. *a.* Seed.

# WOOLLY TIDESTROMIA

PIGWEED FAMILY — Amaranthaceae

### WOOLLY TIDESTROMIA — *Tidestromia lanuginosa* (Nutt.) Standl.

DESCRIPTION — A brittle stemmed summer annual, reproducing by seeds. The whitish stems often are bright scarlet in the fall. They are erect, prostrate, or drooping on the ground, and many-branched. Usually radiating from the root, they are 3 inches to 1½ feet high or long, and may spread to 5 feet in diameter either on the ground or above. The densely hairy leaves are opposite, football or oval shaped, ⅓ to 2 inches long, and ¼ to 1 inch broad. The 2 opposite leaves are usually of different sizes.

The small yellowish flowers, about ⅛ inch long, are borne in a few flowered clusters on the leaf axils. They have 5 stiff papery flower parts and 5 stamens. The small seeds (fruits) are almost globeshaped.

DISTRIBUTION — Woolly tidestromia grows on rocky or sandy soil. It is widespread in Arizona, and is conspicuous after the start of the summer rains, forming whitish mats on mesas, sandy plains, desert slopes, and juniper flats. In southern Arizona, this native plant has invaded the newly irrigated lands, and has become a pest in cotton fields, as in Avra Valley and Pima County, and in other cultivated crops and irrigation ditches; 100 to 6,000 (mostly lower) feet elevation; flowering July to October. It is also one of the hosts of the beet leafhopper.

122

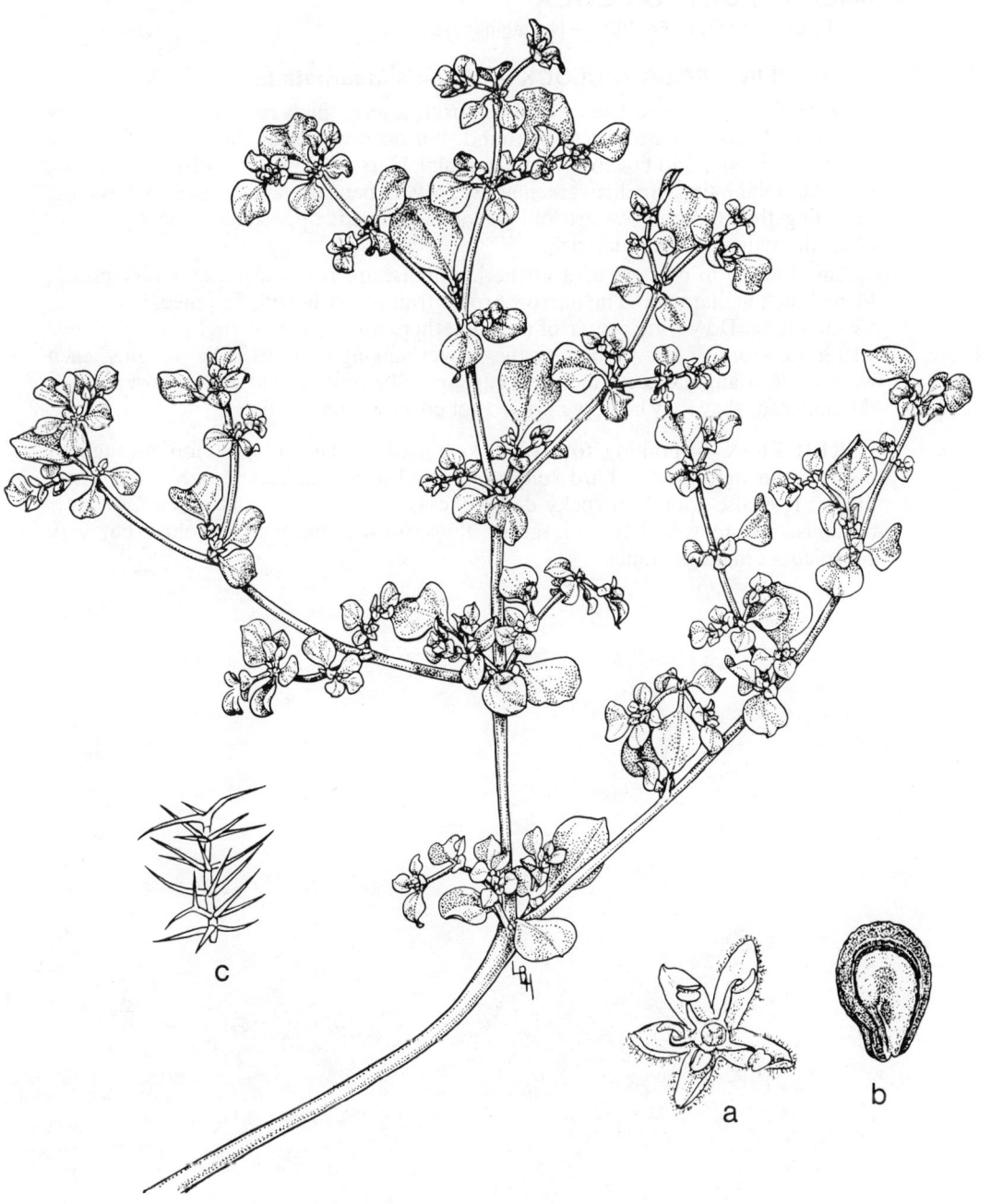

Fig. 57. Woolly tidestromia *(Tidestromia lanuginosa)*.
Flowering branch. *a.* Flower. *b.* Seed. *c.* Branched hair.

123

# TRAILING FOUR O'CLOCK

FOUR O'CLOCK FAMILY — Nyctaginaceae

## TRAILING FOUR O'CLOCK — *Allionia incarnata* L.

DESCRIPTION — A prostrate perennial from a long thick root, reproducing only by seeds. The stems trail on the ground, but do not root at the joints. They are ½ to 3 feet long, and covered with glandular hairs which collect dirt. The leaves, also glandular hairy, are dirty green above, silvery beneath, and ½ to 3 inches long, including the stalks. They are oblong or eggshaped, opposite on the stem, and often the pair is of unequal size.

The flowers, in clusters of 3 at the leaf axils, are rose purple and very pretty, ½ to 1 inch in diameter. The narrow brown fruit is rodshaped, flattened, and about ⅛ inch long. Down the center of one side there are 2 rows of sticky green glands which are more or less hidden by the 2 overhanging margins. The margins, each with 4 to 6 triangular teeth along its edge, may be quite broad and almost meet in the center, or they may be very narrow, not covering the glands.

DISTRIBUTION — Trailing four o'clock, a native weed, is common throughout most of the state on dry hard soil, along roadsides, sidewalks, paths, and waste places. It is also found on rocky desert mesas and slopes, sandy washes, or river bottoms; 100 to 6,000 feet elevation; flowering February to October, but very conspicuous in late summer.

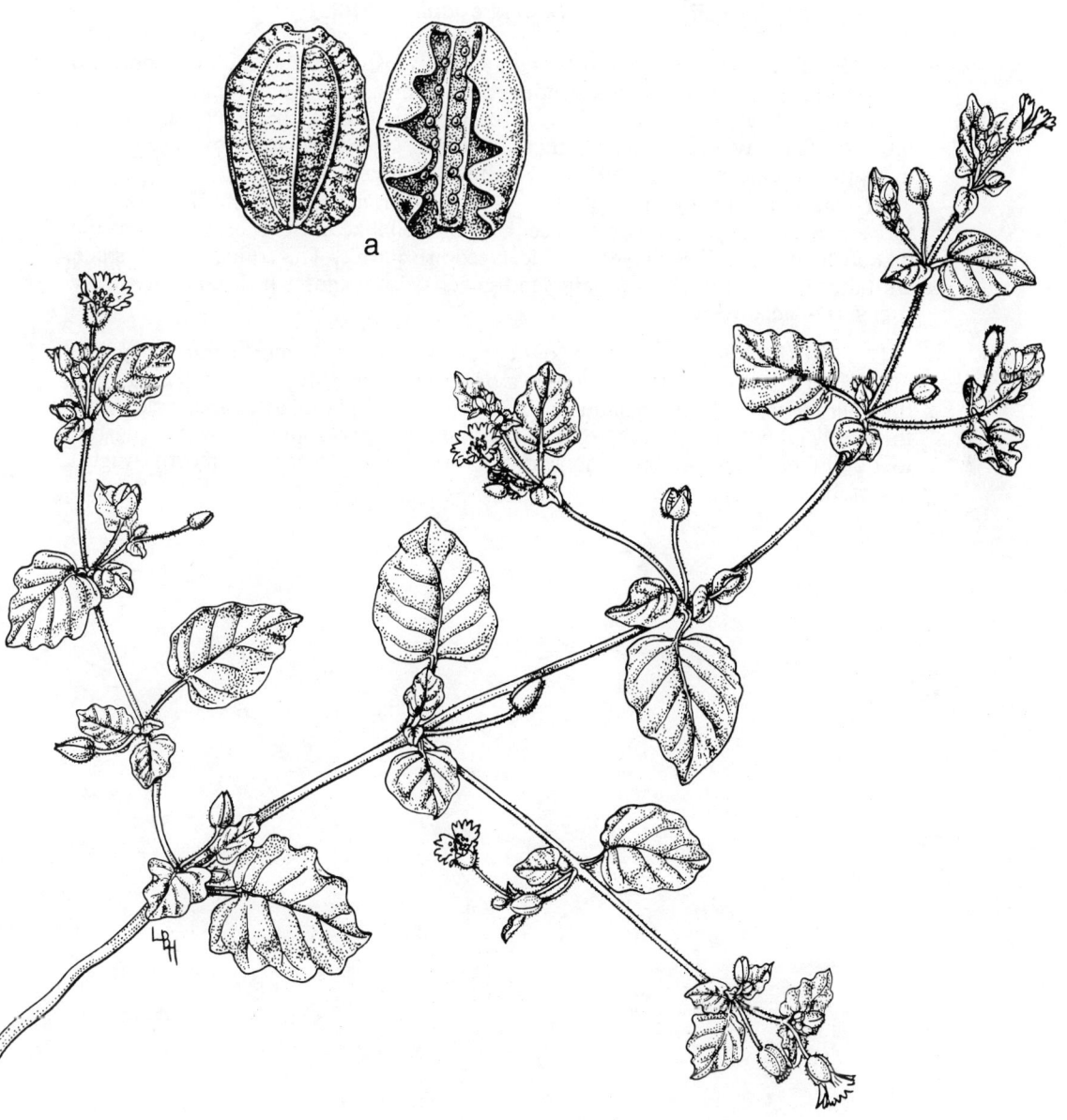

Fig. 58.  Trailing four o'clock (*Allionia incarnata*).
Flowering branch.  *a.* Upper and lower sides of fruit.

## RED SPIDERLING

FOUR O'CLOCK FAMILY — Nyctaginaceae

**RED SPIDERLING — *Boerhaavia coccinea* Mill.**

DESCRIPTION — A stout perennial with tough, prostrate stems radiating outward from a thick woody root and ascending at the ends, 1 to 6 feet long. The stems, often with sticky yellow bands above, are usually noticeably hairy, especially near the base. The leaves are similar to those of coulter spiderling, and are densely hairy to hairless. Some plants, however, are almost hairless.

In maturity, the flowering part is copiously branched with myriads of tiny threadlike stems, and the flowers occur in umbrellalike clusters of 3 to 25 at the tips of these stems. The flowers are deep reddish purple. The fruits, also 5-ribbed and about ⅛ inch long, are hairy and covered by glands which contain a very sticky substance.

DISTRIBUTION — This and coulter spiderlings are the most common spiderlings in Arizona, but red spiderling is more troublesome since it is perennial, and thus harder to eradicate. Found in the same type of places as coulter spiderling, this weed grows from Greenlee, Yavapai, and Mohave counties southward; 130 to 7,000 feet elevation; flowering May to November or December; mostly August to October.

126

Fig. 59. Red spiderling *(Boerhaavia coccinea)*. Basal portion of plant and flowering branch. *a.* Flower cluster, umbrellalike. *b.* Gland-covered fruit.

## COULTER SPIDERLING, coulter boerhaavia
FOUR O'CLOCK FAMILY — Nyctaginaceae

### COULTER SPIDERLING — *Boerhaavia coulteri* (Hook. f.) Wats.

DESCRIPTION — An annual reproducing by seeds. The stems are erect or ascending, 1 to 2½ feet high, or prostrate, radiating outward from the roots and ascending at the ends, 1 to 5 feet long. In rich soil, this plant may cover an area of 10 feet. The lower part of the stems may be very hairy, rarely glandular hairy, or covered by minute hairs. Sticky yellow bands on the upper stems are common particularly at maturity, but often none are present. The opposite leaves are unequal in size, eggshaped, usually hairless except along the stalks and sometimes the edges, and 1 to 4 inches long.

The flowers are pale pink, about 1/12 inch long, in clusters of 1 to 3, and scattered along the upper stem branches for 1 to 2 inches. The wedgeshaped fruits are hairless, with 5 broad ridges, and about ⅛ inch long.

DISTRIBUTION — Found throughout most of the state except in the northeastern part. Especially common and annoying in late summer in southern Arizona in cultivated crops, gardens, roadsides, vacant lots, and overgrazed eroded ranges. Growing also in sandy washes, desert plains, and mountain foothills and slopes; 100 to 5,000 feet elevation; flowering June to October or November.

Fig. 60.  Coulter spiderling *(Boerhaavia coulteri)*. Plant and
enlarged branch in flower and fruit.  *a.* Flower.  *b.* Fruit.

129

# HORSE PURSLANE
CARPETWEED FAMILY — Aizoaceae

## HORSE PURSLANE — *Trianthema portulacastrum* L.

DESCRIPTION — Horse purslane is a fleshy succulent annual with a shallow root, reproducing only by seeds. At first the weak diffusely branched stems are erect, but seldom higher than 1 foot (or to 2 or more feet in competition for light, as in a soybean field) before spreading, and finally reclining on the ground, 1 to 5 feet long. The bright green fleshy oval leaves are opposite, the pairs very unequal in size, ½ to 3½ inches long including the stalks which are widened and 2-toothed at the base, and are ⅓ to 1⅔ inches broad.

The flowers have 6 to 10 stamens, and 5 flower parts about ⅓ inch long. These flowers are rose purple within, and each bears a thickened hornlike tip on the back. The flowers occur singly or a few together in the leaf and branch axils, and are open only in the mornings. The small seedpod is topshaped, with the upper part thickened, 2-crested and falling away as a lid when the 1 to 5 seeds are mature. The thick blackish seeds are squarish or somewhat heartshaped, about 1/12 inch broad, and covered by whitish wavy ridges.

DISTRIBUTION — Horse purslane is a tropical and subtropical American plant. It is a troublesome annual weed common in sorghum, citrus, and cotton; up to 4,000 feet elevation. It is also common in alkaline flats, roadsides, lawns, and waste places; flowering May to November. This plant is a host for the beet leafhopper.

Fig. 61. Horse purslane *(Trianthema portulacastrum)*.
Flowering plant. *a.* Portion of flowering branch. *b.* Flower
enlarged. *c.* Fruit two-crested, with seeds. *d.* Seed.

131

## COMMON PURSLANE, pursley, pusley, wild portulaca
PURSLANE FAMILY — Portulacaceae

### COMMON PURSLANE — *Portulaca oleracea* L.

DESCRIPTION — A smooth fleshy annual reproducing by seeds, closely resembling horse purslane, with which it is often confused. The many branched stems are reddish and prostrate, ½ to 2 feet long, and often form mats with the tips turned upward. Sometimes the stems are ascending and nearly erect. The small thick leaves are alternate, not opposite as in horse purslane, either solitary or clustered, and spatulashaped with the tips rounded.

The small stalkless flowers, as in horse purslane, occur singly or several together in the leaf and branch axils and stem tips. They also open only in the mornings, but are yellow, not purplish. There are 7 to 12 or 20 stamens.

The seed pod is globeshaped, the upper half of which, with the 2-cleft calyx on top, falls away as a lid when the many tiny seeds are mature. The black seeds, broadly eggshaped but flattened, are less than 1/25 inch long with a white spot at the scar.

DISTRIBUTION — Introduced from Europe, common purslane is abundant in northern as well as southern Arizona. It is a pest in lettuce, sugar beets, carrots, and citrus; also common on overgrazed eroded areas, on mountain slopes and meadows; 100 to 8,500 feet elevation; flowering April to June and August to about November.

Fig. 62. Common purslane *(Portulaca oleracea).* Flowering plant. *a.* Flower. *b.* Seed.

133

# BARESTEM LARKSPUR
CROWFOOT FAMILY — Ranunculaceae

### BARESTEM LARKSPUR — *Delphinium scaposum* Greene

DESCRIPTION — A grayish green perennial which reproduces by seeds and from clusters of dark hard woody roots, often forming large or small colonies. The stems are solitary or few, ½ to 2½ feet high. The leaves are divided into narrow segments, and mostly clustered at the base.

The flowers are usually an intense royal blue, but are sometimes lighter blue or nearly lavender. They look exactly like the cultivated larkspurs, with the characteristic long spur at the base. The seedpods, ⅜ to ¾ inch long, split into 3 parts when the many seeds are mature.

DISTRIBUTION — Barestem larkspur is the most widespread larkspur in Arizona. Found in sandy or gravelly soil on desert mesas and foothills in Apache, Navajo, Coconino, Mohave, Yavapai, Gila, Maricopa, Pinal, Graham, and Pima counties; 2,000 to 5,000 (rarely 8,000) feet elevation; flowering March to April, or May and early June in the higher elevations.

Larkspurs, where sufficiently abundant in Arizona, are a serious cause of livestock poisoning. Cattle losses from larkspur poisoning occur very early in the spring. Santa Cruz County, particularly in the Sonoita area, Yavapai County, in the vicinity of Prescott, and Coconino County, near Seligman and between Williams and Flagstaff are known trouble areas.

The 9 native Arizona larkspurs may be divided into tall larkspur (2½ to 6 feet tall, of high moist elevations, flowering in late summer) and low larkspur (½ to 2 feet tall, of dryer plains and foothills, flowering in the spring). All species should be regarded as poisonous. The tall larkspurs may be more poisonous, but the low larkspurs are more abundant, and probably more responsible for the cattle poisoning.

### WOOTON PLAINS LARKSPUR — *Delphinium virescens* Nutt. subsp. *wootoni* (Rydb.) Ewan

Very similar to barestem larkspur, but the leaves and stems are covered by fine incurved white hairs. Sometimes a few leaves are found just above the basal cluster. The flowers are whitish or pale lavender blue, fading buff, and the lower petals are white with conspicuous white hairs.

Found in southeastern Arizona on open slopes and plains in Graham, Cochise, Santa Cruz, and Pima counties; 3,800 to 6,000 feet high; flowering April and May. Wooton plains larkspur is probably the cause of larkspur poisoning in Santa Cruz County.

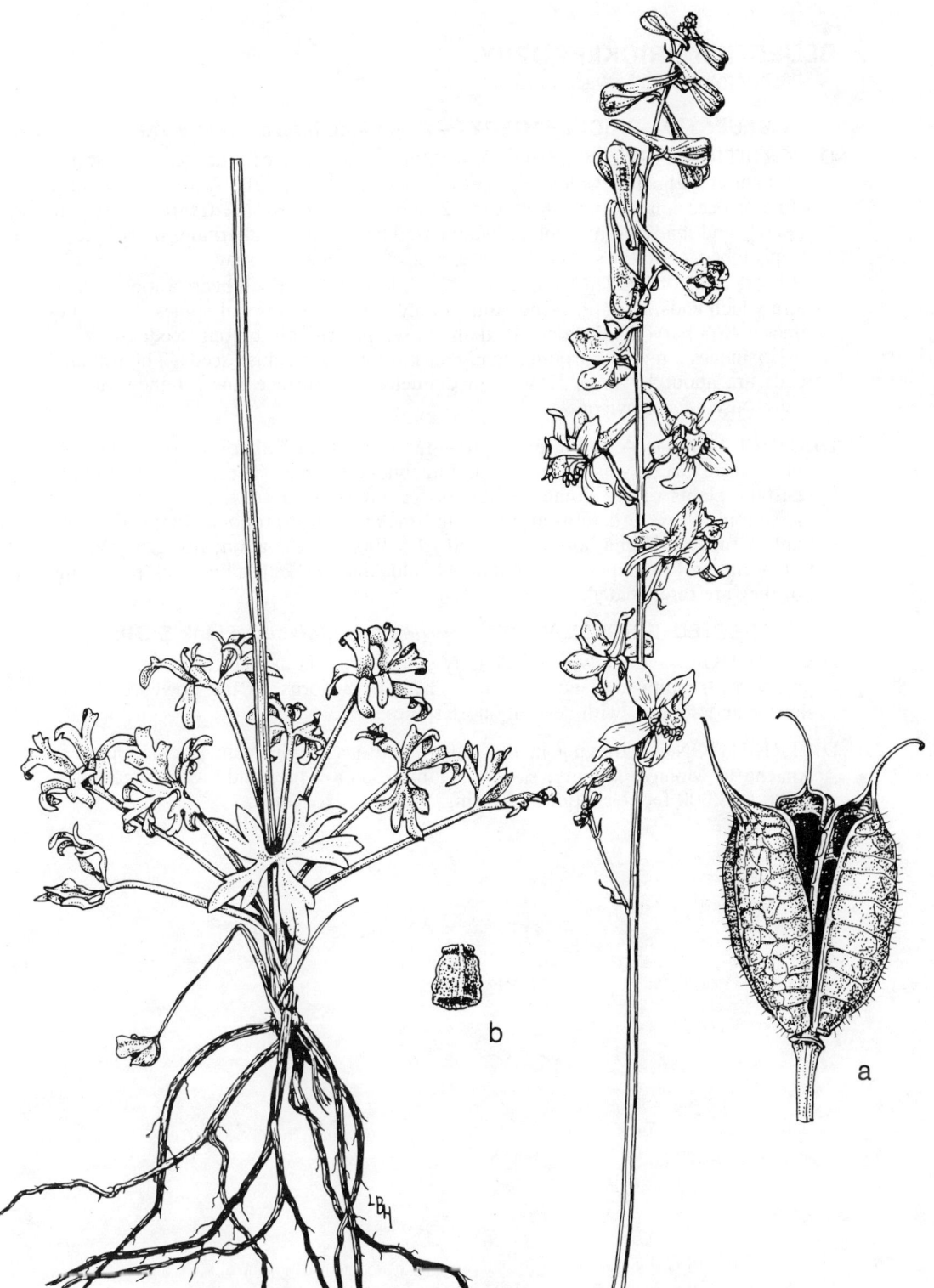

Fig. 63.  Barestem larkspur *(Delphinium scaposum)*. Basal portion of
plant and flowerstalk with spurred flowers.  *a.* Seed pod.  *b.* Seed, angled.

# BLUESTEM PRICKLEPOPPY

POPPY FAMILY — Papaveraceae

## BLUESTEM PRICKLEPOPPY — *Argemone intermedia* Sweet

DESCRIPTION — Prickly perennials with bitter yellow juice, and densely covered throughout with short yellowish spines, which reproduce only by seeds. The large white or occasionally pinkish flowers, 2 to 3 inches across, are fragrant with 4 to 6 petals and many orange colored stamens. The leaves are alternate, bluish green, deeply lobed, clasp the stem at the base, and are 2 to 8 inches long.

There are 2 or 3 green sepals covering the flower bud. Each bears a long slender horn which ends in 1 stiff spine, and usually with no additional spines, but sometimes 1 to 3 very slender ones near the base. The prickly oblong seedpods are 1 to 1½ inches long, and produce many dark brown or blackish seeds. The rounded seeds are about ⅙ to 1/12 inch in diameter, the surface finely honeycombed, with a raised scar down one side.

DISTRIBUTION — A native plant growing in dry disturbed soil of roadsides, old fields, waste places, washes, mesas, and uncultivated areas. These are very drought resistant plants. They come in abundance on overgrazed ranges, and are an indication of severe deterioration. Found in Yavapai, Maricopa, Pinal, Cochise, Santa Cruz, and Pima counties; 1,500 to 5,400 feet elevation; flowering March to November. Pricklepoppies contain alkaloids that may cause livestock poisoning, but they are rarely eaten.

## CRESTED PRICKLEPOPPY — *Argemone platyceras* Link & Otto

DESCRIPTION — Usually more densely spiny throughout than bluestem pricklepoppy, with many fine short bristles. The 2 or 3 horns on the flower buds are shorter and stouter, with several short spines.

DISTRIBUTION — Growing in the same situations as bluestem pricklepoppy, in Apache to Mohave County, south to Pima, Santa Cruz, and Cochise counties; 1,500 to 8,000 feet elevation; flowering March to November.

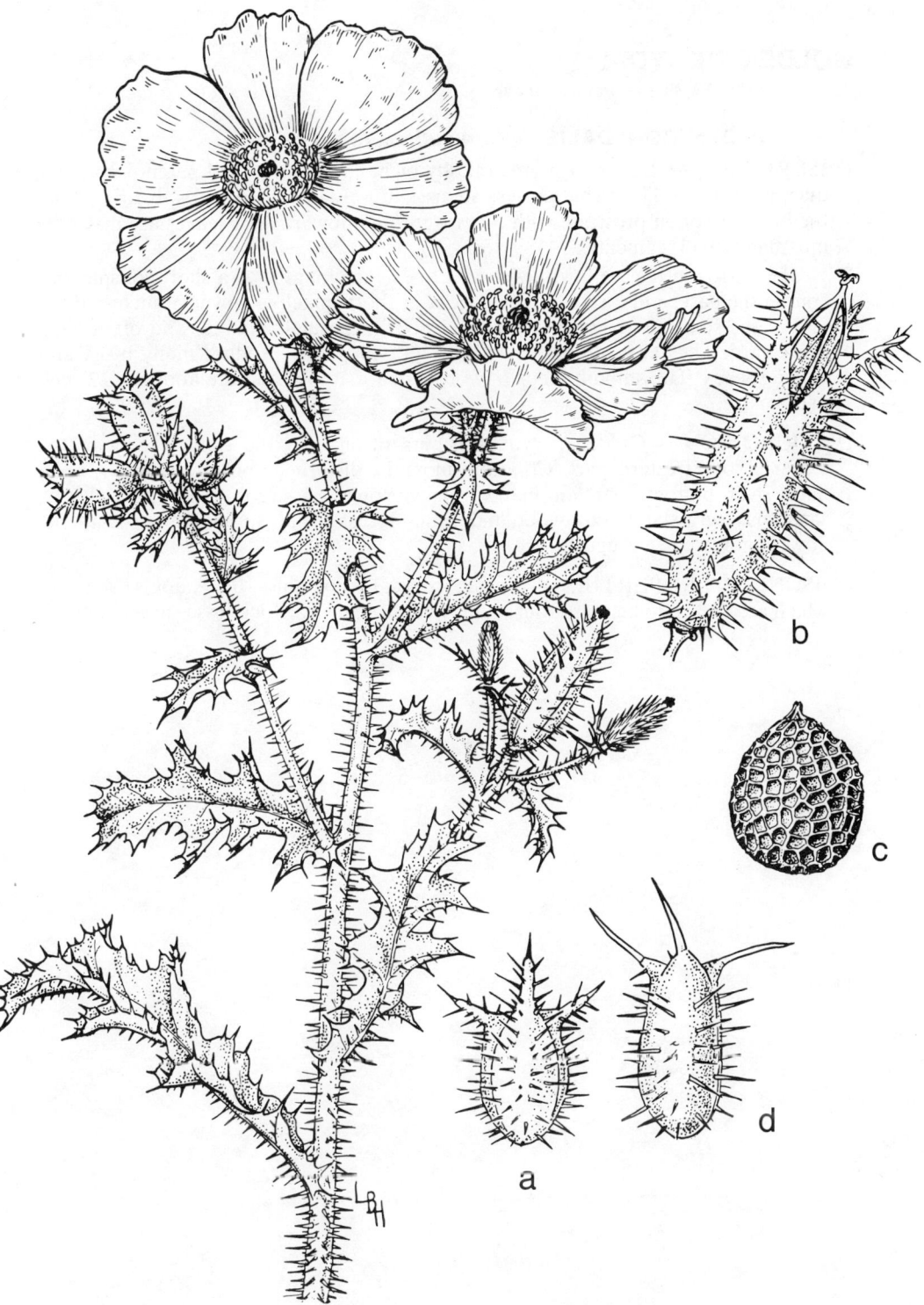

Fig. 64. Crested pricklepoppy *(Argemone platyceras)*. Branch with
two large flowers. *a.* Flower bud with three spiny terminal horns.
*b.* Seedpod. *c.* Seed. *d.* Bluestem pricklepoppy *(A. intermedia)*.
Flower bud with terminal horns lacking spines, except at base.

137

# GOLDEN CORYDALIS

POPPY FAMILY — Papaveraceae

## GOLDEN CORYDALIS — *Corydalis aurea* Willd.

DESCRIPTION — Low bluish green fleshy biennial or short lived perennial, reproducing by seeds. The weak hairless stems, diffusely branched and spreading from the base, are often prostrate with the tips turning upward. The leaves are dissected into many small segments.

The narrow bright yellow flowers, ⅜ to ¾ inch long, have a short conspicuous spur at the base, and 2 sepals. They occur in short spikelike flower branches at the end of the stems. The bluish green seed pods are usually curved, and often hang downward. They are cylindrical, ½ to 1 inch long, and produce many black and shining seeds. The smooth seeds have rounded margins, and are about 1/12 inch in diameter.

DISTRIBUTION — Golden corydalis occurs throughout the state, except in the extreme southwestern part. Common in moist disturbed soil of roadsides, open flats, river bottoms, stream banks, rocky slopes, in creosote-mesquite desert, sagebrush, grassland, oak-woodland, pine forests, and spruce-fir associations; flowering March to August.

POISONOUS PROPERTIES — Golden corydalis contains 10 alkaloids, some of which are known to be poisonous. It may cause some livestock losses in Arizona.

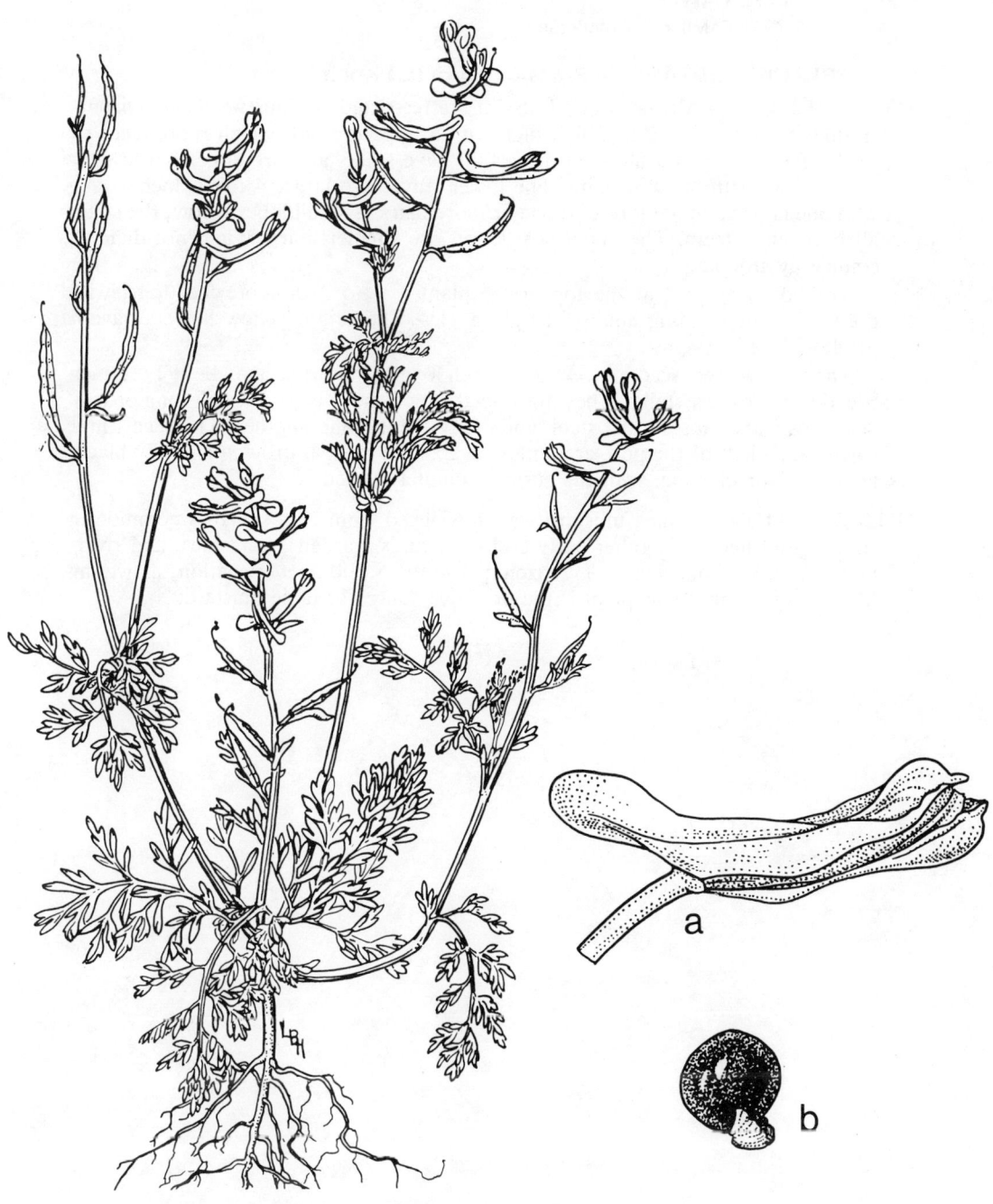

Fig. 65. Golden corydalis *(Corydalis aurea)*. Flowering plant.
*a.* Flower with conspicuous spur at base. *b.* Seed.

# BLACK MUSTARD

MUSTARD FAMILY — Cruciferae

## BLACK MUSTARD — *Brassica nigra* (L.) Koch.

DESCRIPTION — All species of *Brassica* are restricted noxious weeds in Arizona. A stout erect annual, 2 to 6 feet high with a large taproot, which reproduces by seeds. The dark green alternate leaves all have stalks and are either hairless, or have a few scattered stiff hairs. The lower leaves are large, 4 to 16 inches long with one large terminal lobe and one or more pairs of small lobes below; the edges all have small teeth. The uppermost leaves are smaller, narrow and not divided, commonly drooping.

The flowering part at the top of the plant is 1 to 2 or more feet long, with the branches spreading almost at right angles. The bright yellow flowers have 4 petals with yellow veins.

The narrow erect seedpods, ⅜ to ¾ inch long at maturity, and about 1/12 inch broad, are on short stalks. They are tipped by a slender beak ⅛ inch long or less. The pods stand close to the stem, and often overlap one another. The seeds, in 1 row in each half of the pod, are almost oval, dark reddish brown or nearly black, about 1/16 inch long, with the surface minutely pitted.

DISTRIBUTION — Black mustard was introduced from Eurasia. It is common in spring grain fields and other early crops, pastures, gardens, roadsides, and river-bottoms throughout most of Arizona; 100 to 8,300 feet elevation; flowering March to October. This plant is the principal source of table mustard.

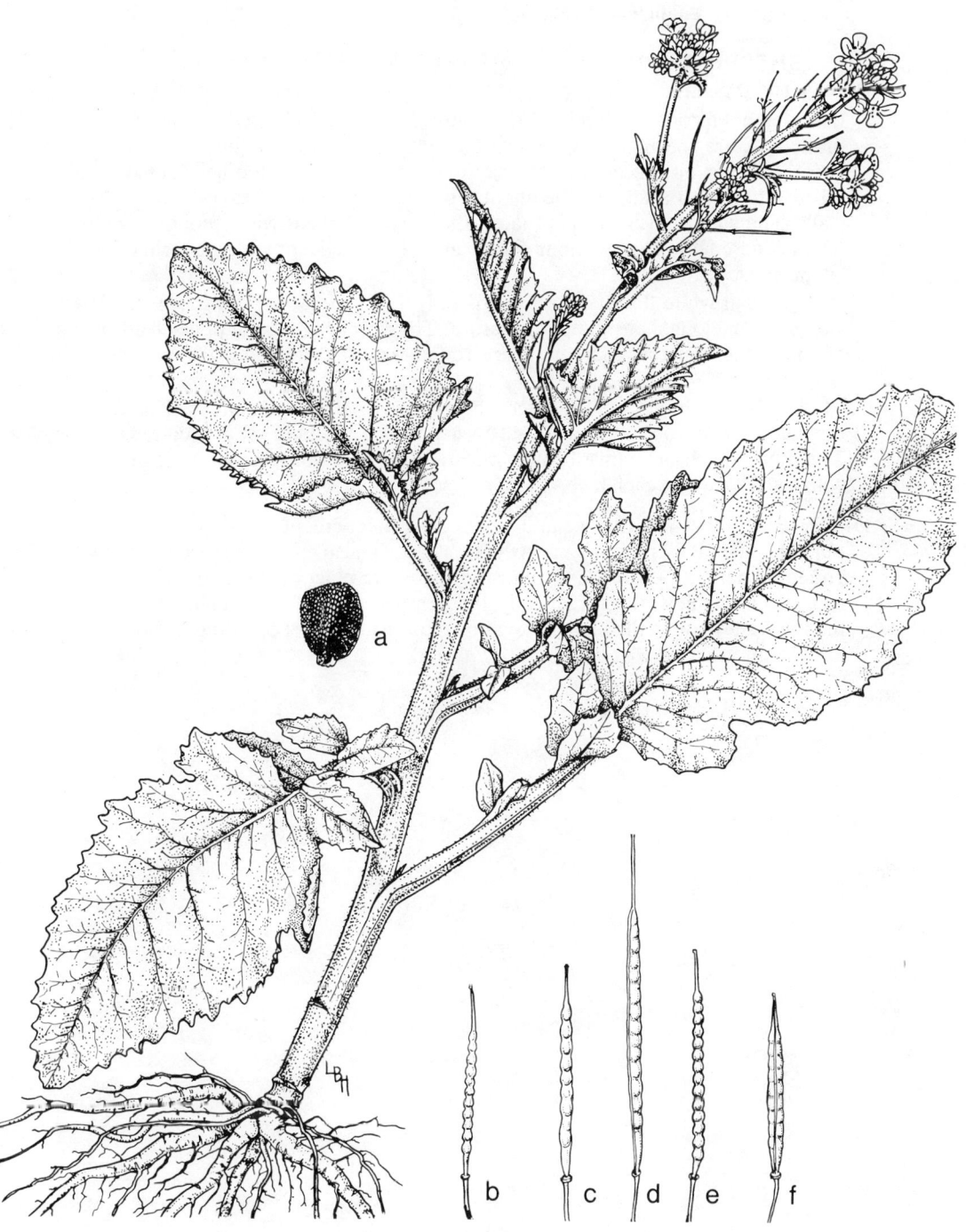

Fig. 66.  Black mustard *(Brassica nigra)*. Flowering plant.  a. Seed.  b. Seedpod.
c. Indian mustard *(B. juncea)*. Seedpod.  d. Wild turnip  *(B. campestris)*. Seedpod.
e. Wild mustard, charlock *(B. kaber)*. Seedpod.  f. Rocketsalad, roquette *(B. eruca)*. Seedpod.     141

# SHEPHERDSPURSE

MUSTARD FAMILY — Cruciferae

## SHEPHERDSPURSE — *Capsella bursa-pastoris* (L.) Medic.

DESCRIPTION — An erect annual or winter annual, ¼ to 1½ feet high, which has a thin taproot, and reproduces only by seeds. The slender stems, usually branching, are sparingly covered by long gray hairs. A spreading rosette of leaves is formed first on the ground. These leaves are variously toothed, cut or deeply lobed, often with a larger lobe at the tip, and 1½ to 5 inches long including the stalk. The stem leaves are alternate, often arrowshaped, with smooth or toothed edges. These leaves are without stalks, and clasp the stem with an earlike lobe on either side.

The small white flowers, with petals only 1/12 to ⅛ inch long, are on slender stalks which elongate as the pods mature. They occur along the upper leafless part of the stems, but the pods often are found almost throughout the length of the stem.

The flat seedpods are inverted, heartshaped, or triangular, with the broad notched end on top, and narrowed to a point at the base, about ¼ inch broad. The tiny reddish or orange brown seeds are oblong, shiny, about 1/25 inch long, with a groove down each face.

DISTRIBUTION — A European introduction, shepherdspurse is common throughout the state. It is primarily a lawn pest, but is also very common in cultivated crops, orchards, pastures, and roadsides or waste places near them; 100 to 8,000 feet elevation; flowering practically the year around in moist cultivated fields in the low valleys or at the higher elevations, but usually disappearing in May in dry, low elevation areas.

Fig. 67.  Shepherdspurse *(Capsella bursa-pastoris)*. Plant with flowers and fruits and two enlarged leaves showing variation. *a.* Flower. *b.* Seedpod, enlarged. *c.* Seed.

143

# HOARY CRESS, whitetop
MUSTARD FAMILY — Cruciferae

### HOARY CRESS — *Cardaria draba* (L.) Desv. (*Lepidium draba* L.)

DESCRIPTION — A prohibited noxious weed in Arizona, hoary cress is a grayish green, minutely hairy perennial, ½ to 2 feet high, which reproduces by seed, but mostly by thick deeply penetrating vertical and wide spreading, creeping branch roots. The stems are erect and greatly branched in the flowering part. The broadly elliptical leaves, 1 to 3½ inches long, have toothed to nearly smooth margins. Only the lower leaves are stalked; the others are stalkless and clasp the stem with 2 earlike lobes.

The short flowering branches bear many very small flowers, each with 4 petals about ⅛ inch long, giving the plant a flattop appearance. The flowers and the small seedpods are borne on slender spreading stalks about ¼ to ½ inch long. The mature seedpods are either heartshaped and broadest at base, or broader above and rounded to obtuse at the base. They are nearly hairless, ⅛ to ¼ inch broad, and tipped by a slender persistent beak. A central partition separates the 2 halves, each producing 1 (rarely 2) red-brown seed about 1/16 inch long, and about 1/16 inch broad.

DISTRIBUTION — Hoary cress, introduced from Europe and Western Asia, forms large patches in cultivated fields, gardens, pastures, and roadsides. Locally common on ranches in the Springerville-Eager area (Apache County) to Peeples Valley (Yavapai County), northward to Fredonia (Coconino County), and occasionally elsewhere in Arizona; 3,500 to 7,000 feet elevation; flowering April to July. Not widespread enough yet in Arizona to be a major problem, but it should be eradicated where found to prevent its spreading.

### HAIRY WHITETOP, hairy-podded whitetop — *Cardaria pubescens* (C. A. Mey.) Roll. var. *elongata* Roll. (*Hymenophysa pubescens* auth.)

DESCRIPTION — Unknown to date in Arizona, but is placed on the prohibited noxious-weed seed list to prevent its entry through commercial seed. Often difficult to distinguish from hoary cress, but differing in its closely hairy seedpods, which are slightly smaller and nearly globeshaped.

144

Fig. 68. Hoary cress *(Cardaria draba)*. Flowering plant showing horizontal stems. *a.* Basal leaf. *b.* Seedpod. *c.* Seed.

# TANSYMUSTARD

MUSTARD FAMILY — Cruciferae

## TANSYMUSTARD — *Descurainia pinnata* (Walt.) Britt.

DESCRIPTION — An erect annual or winter annual, ½ to 3 feet high, which reproduces only by seeds. The plants are usually ash colored from the many short soft, forked or starlike hairs, or they may be only moderately hairy and green. The stems are often purplish. The alternate leaves are divided once or twice into fine segments, 1 to 4 inches long.

The tiny flowers, about 1/12 inch long, are yellow or whitish, and on short slender spreading stalks, which elongate as the pods mature. They occur along the upper part of the stems. The short narrow seedpods are somewhat stubby, ¼ to ½ inch long, and have 2 rows of seeds in each half. The tiny oblong seeds are dull red, about 1/25 inch long, and have a groove on 1 side.

DISTRIBUTION — Tansymustard is a native weed which is abundant in moist spots throughout the state on sandy, rock, or disturbed soil of riverbeds, washes, mesas, canyons, slopes, and swales in creosote desert, grassland, sagebrush, oak, and pinyon-juniper associations. It is also a troublesome weed in cultivated lands, grain fields, city streets, and waste places; 100 to 7,000 feet elevation; flowering December to August, mostly March and April at the lower elevations.

POISONOUS PROPERTIES — Tansymustard contains no known poisonous principles, but may cause trouble under certain conditions. These plants appear on the ranges in abundance in early spring, and livestock may eat them almost exclusively. Animals may be unable to eat or drink, as a result of eating large quantities over a long period of time. The pods are also relished by livestock. The mature seeds are rich in oil of mustard.

## FLIXWEED — *Descurainia sophia* (L.) Wats.

DESCRIPTION — Similar to tansymustard, but with much more finely divided leaves, brighter green and less hairy. The pods are much longer, ¾ to 1⅓ inch long, and very narrow; the seeds are in 1 row in each half.

DISTRIBUTION — Introduced from Eurasia, flixweed is found in the same type of places with the same general distribution as tansymustard, but is not as abundant except in some local areas; flowering mostly March to May.

Fig. 69.  Tansymustard *(Descurainia pinnata).* Flowering plant.
*a.* View of portion of seedpod showing seed attachment.  *b.* Seedpod.  *c.* Seed.

147

# SAND PEPPERGRASS
MUSTARD FAMILY — Cruciferae

## SAND PEPPERGRASS — *Lepidium lasiocarpum* Nutt.

DESCRIPTION — Bushy annual, 1 inch to 1¼ feet high, reproducing only by seeds. The stems are erect or prostrate, usually profusely branched from the base, with short stiff straight hairs lying flat on the surface, or with rather long stiff bristly spreading hairs. The leaves, 1 to 6 inches long including the stalks, are alternate, broader toward the tip, and may be merely toothed, deeply lobed, or cut into very fine segments. They often drop off as the plant matures.

The petals are tiny and white, less than 1/12 inch long, but are often lacking completely. The rounded seedpods, notched at the tip, are covered with short bristly hairs, or sometimes are hairy only along the edges, ⅛ to ⅙ inch long, and about ⅛ inch broad. Their stalks are not rounded, but noticeably flattened, hairy on both sides, or hairless on the lower surface. They contain only 2 seeds, 1 in each half of the pod. The reddish oval seeds have a membranous margin, narrow all around the edge, but wider at the top.

DISTRIBUTION — Sand peppergrass, a native plant, is abundant throughout the state in the northern as well as the southern part, in sandy or rocky disturbed soil along roadsides, in fields, pastures, waste places, eroded hillsides on overgrazed ranges, mesas, washes, and river bottoms; 100 to 6,500 feet elevation; flowering January to June, but mostly March and April.

Fig. 70. Sand peppergrass *(Lepidium lasiocarpum).*
Flowering plant. *a.* Flower. *b.* Seedpod. *c.* Seed.

149

# GORDON BLADDERPOD

MUSTARD FAMILY — Cruciferae

## GORDON BLADDERPOD — *Lesquerella gordoni* (Gray) Wats.

DESCRIPTION — A low bushy, erect, spreading or prostrate annual or winter annual, ¼ to 2 feet high, reproducing only by seeds. The narrow lanceshaped or spatulashaped leaves are 1/12 to ½ inch broad, mostly smooth edged, sometimes toothed; the basal leaves may be lobed. The grayish starlike hairs are often so dense as to give the plant a silvery appearance.

The bright yellow flowers are ¼ to ⅓ inch long, often fading reddish. The inflated ballshaped pods, ⅛ to ⅕ inch in diameter, are tipped by a slender point (the persistent style) 1/12 to ⅛ inch long, and are hairy or hairless. The stalks of the mature pods are spreading, often curved twice in opposite directions like the letter "S". The reddish brown, flattened fanshaped seeds are about 1/12 inch broad, and 1/16 inch long, with crinkled edges. There are several seeds in each half of the pod.

DISTRIBUTION — An abundant native plant, Gordon bladderpod is a weed in irrigated fields, roadsides, yards, disturbed soil, and overgrazed desert ranges. It also grows in river bottoms, mesas, plains, slopes, and canyons throughout southern Arizona from Greenlee to Mohave counties and southward; 110 to 5,000 feet elevation; flowering February to May.

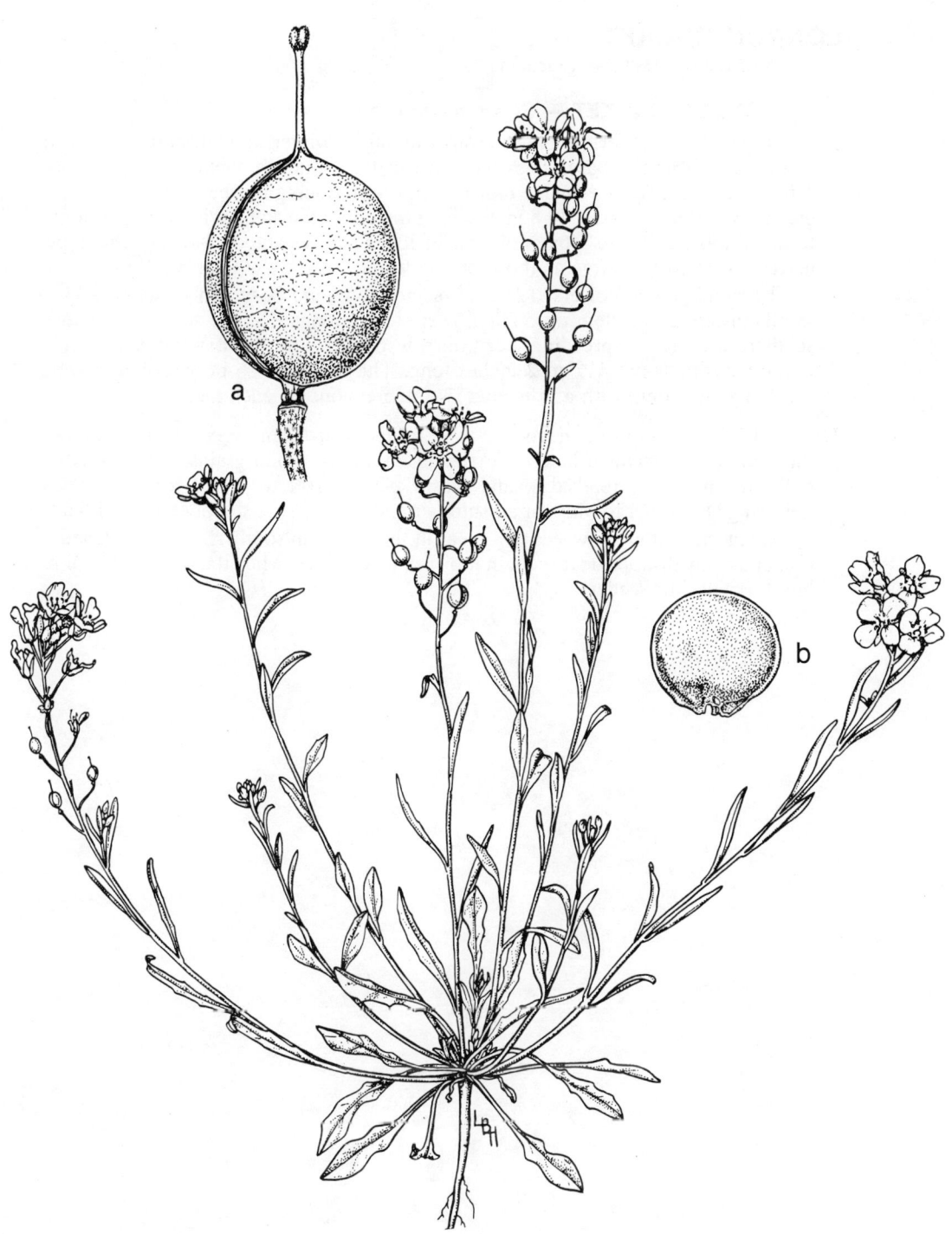

Fig. 71. Gordon bladderpod (*Lesquerella gordoni*). Plant with flowers and fruits. *a*. Fruit or seedpod. *b*. Seed.

# LONDON ROCKET

MUSTARD FAMILY — Cruciferae

### LONDON ROCKET — *Sisymbrium irio* L.

DESCRIPTION — A bright green fleshy annual or winter annual, hairless or with a few hairs near the base. The stems are usually much branched from the base 1 to 3 feet high, from a coarse taproot, and reproducing by seeds only. The dark green divided leaves are 1 to 8 inches long including the stalk, with a large pointed terminal lobe, and 1 to 4 pairs of smaller lobes below. The divisions of the upper leaves are almost as large as those on the lower.

The small yellow flowers, 1/12 to ⅛ inch long, are borne on slender stalks in small clusters at the stem tips. The flower stems elongate as the seedpods mature, so there are many, spreading, very narrow pods below the flower clusters. The mature seedpods are 1½ to 2 inches long. The tiny reddish brown oblong seeds are 1/16 inch long, with a ridge and two depressions on each face.

DISTRIBUTION — London rocket is a native European weed. It is abundant throughout the irrigated lands of Arizona, in alfalfa, small grains, gardens, citrus orchards, pastures, roadsides, and waste places; 100 to 4,500 feet elevations; flowering December to May, or all summer and fall in moist soil of cultivated fields.

One of the first green weeds to appear in the winter in southern Arizona, London rocket usually disappears, except in moist shaded places, when the weather becomes hot. It is a prolific seeder.

Fig. 72. London rocket *(Sisymbrium irio)*. Plant in flower and in
fruit. Also, fruiting branch. *a.* Flower. *b.* Fruit or seedpod. *c.* Seed.

153

# WESTERN CLAMMYWEED, clammyweed

CAPER FAMILY — Capparidaceae

## WESTERN CLAMMYWEED — *Polanisia trachysperma* Torr. & Gray

DESCRIPTION — An erect branching glandular hairy annual, very sticky and clammy to the touch, with an unpleasant odor, ⅓ to 3 feet high, reproducing by seeds. The leaves are divided into 3 lanceshaped leaflets; the upper and those of the flowering stems are not divided.

The 4 petals are whitish or yellowish, but the numerous (12-32) stamens, with their long purple stalks, give the flowers a purplish appearance. The flowers are densely crowded along the upper part of the stems. The sticky seedpods resemble mustard pods. They are long and slender, and at maturity are 1½ to 2½ inches long, on jointed or unjointed stalks ⅞ to 1 inch long. The reddish brown seeds are rounded, about 1/12 inch in diameter, grooved through the center, and the surface is minutely pitted in circular lines.

DISTRIBUTION — Western clammyweed is a native plant growing in disturbed soil along roadsides, waste places, denuded areas, and in sandy canyon washes or stream beds from Navajo and Coconino counties southward; 1,000 to 6,500 feet elevation; flowering May to October.

154

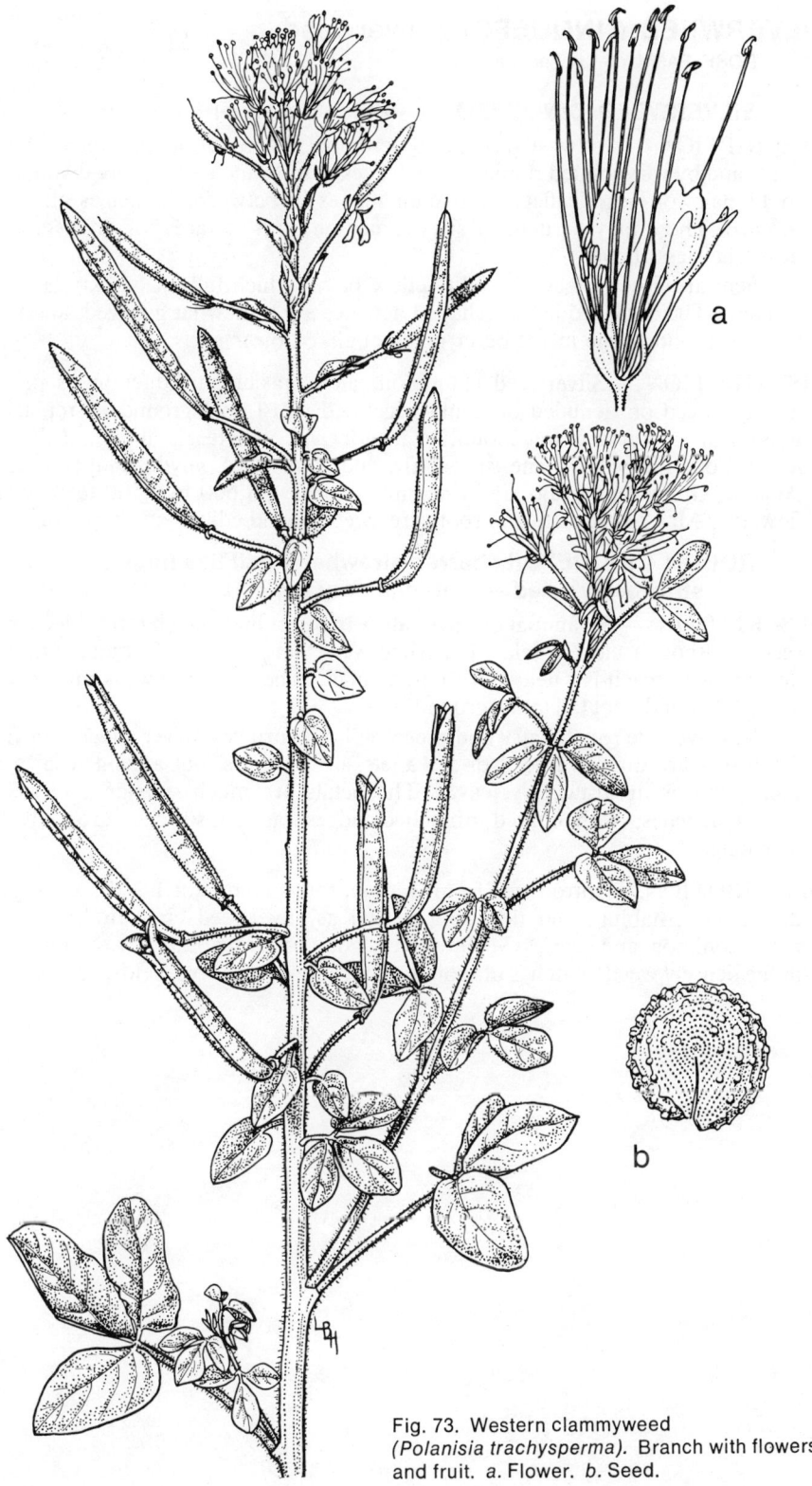

Fig. 73. Western clammyweed
*(Polanisia trachysperma)*. Branch with flowers
and fruit. *a.* Flower. *b.* Seed.

155

## SILVERWEED CINQUEFOIL, silverweed
ROSE FAMILY — Rosaceae

### SILVERWEED CINQUEFOIL — *Potentilla anserina* L.

DESCRIPTION — A low silvery tufted perennial, ¼ to 1 foot high, reproducing by seeds and by wiry jointed runners. The leaves, 2 to 10 inches long, are divided into 5 to 11 pairs of large leaflets with smaller ones in between, the edges all coarsely toothed. These may be densely silky, hairy on both surfaces, or often green and nearly hairless above.

There are 5 conspicuous bright yellow petals which fall off quickly, and many stamens. The thick reddish eggshaped achenes are somewhat grooved, about 1/12 inch long, and borne in the outer green cuplike flower parts.

DISTRIBUTION — Silverweed cinquefoil, an Eurasian introduction, is primarily a range weed on denuded or thinly vegetated moist cool ground. It replaces the grasses on overgrazed wet mountain meadows, and often is abundant in certain areas. Found in yellow pine, spruce-fir, and sometimes pinyon-juniper ranges, in Apache, Greenlee, Navajo, and Coconino counties; 5,600 to 9,500 feet elevation; flowering May to August. The roots are sweetish and edible either raw or cooked.

### ROUGH CINQUEFOIL, barren strawberry, tall five finger, strawberry weed — *Potentilla norvegica* L.

DESCRIPTION — An annual or biennial ½ to 3 feet high which reproduces only by seeds. Although closely related to silverweed cinquefoil, it is quite different in appearance since it is a nearly erect green plant. The stems are weak and spreading at the top, but do not fall to the ground.

The leaves are rough hairy and green on both surfaces, never silvery silky. Also the leaves are divided into 3 or 5 leaflets, not in pairs, but arising from a common point as in strawberry leaves. The petals are much smaller, and the light brown achenes, with curved branched ridges on the surface, are only 1/25 inch long.

DISTRIBUTION — Introduced from Europe, rough cinquefoil has the same general distribution, habitat, and flowering period as silverweed cinquefoil. It is much more common and aggressive, however, competing favorably with other weeds in the densely weedy patches of mountain meadows, pastures, fields, and roadsides.

156

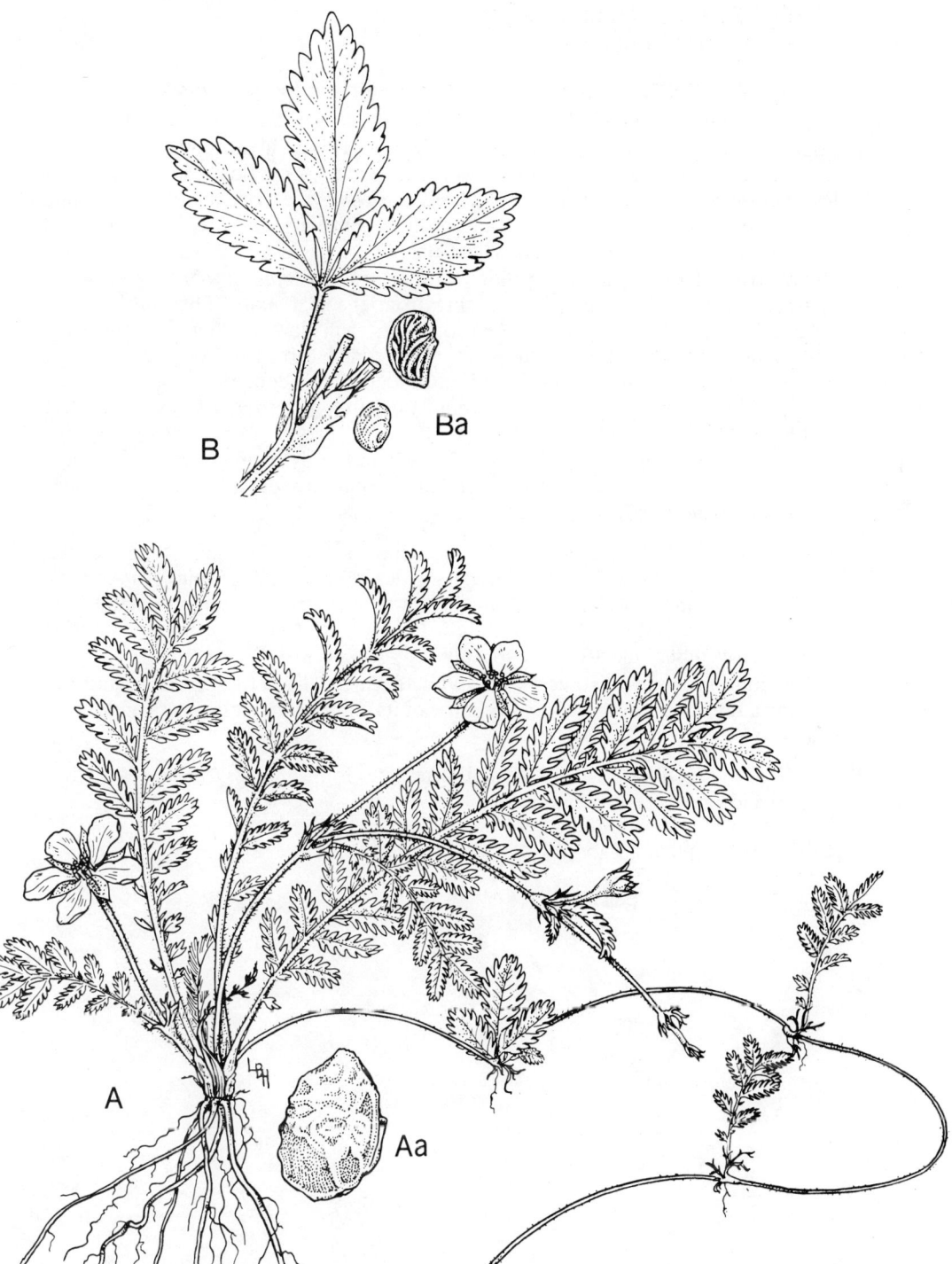

Fig. 74. *A.* Silverweed cinquefoil *(Potentilla anserina).* Flowering plant with runners, the long leaves with 10 to 22 coarsely toothed leaflets. *Aa.* Achene with enclosed seed.
*B.* Rough cinquefoil *(P. norvegica).* Trifoliolate leaf. *Ba.* Two achenes (one enlarged).

157

# WHITETHORN, mescat acacia

PEA FAMILY — Leguminosae

## WHITETHORN — *Acacia constricta* Benth. and var. *vernicosa* (Standl.) Benson

DESCRIPTION — Whitethorn is a tall shrub up to 10 feet high, mostly armed with slender straight white spines in pairs, ¼ to 1½ inches long. Some limbs are spineless, or often entire plants may be spineless. The leaves are twice divided, first into 3 to 7 (or sometimes only 1 or 2) pairs of main divisions, and are further divided into 5 to 16 pairs of tiny secondary leaflets.

The small fragrant golden yellow flowers are clustered into many flowered conspicuous balls on long stalks arising from the leaf axils. They derive their color principally from the many (30 to 40 on each flower) stamens, which are much longer than the 5 tiny yellow petals. The reddish brown curved pods, 2½ to 5 inches long, are about ⅛ inch broad and indented, or constricted between each seed. The boatshaped smooth seeds, about ¼ inch long and 1/12 inch broad, are mottled dark brown, gray, and black.

In southeastern Arizona, the twigs, leaves, and pods of many plants are very glandular and sticky. These are chihuahua whitethorn or stickyleaf whitethorn (*Acacia constricta* var. *vernicosa*).

DISTRIBUTION — Abundant in rocky caliche or limestone soil along washes, slopes, and mesas throughout the deserts and desert grasslands of southern Arizona. It also occurs in central Arizona in the lower chaparral, desert grassland, and desert shrub associations; 2,000 to 5,000 (rarely 6,250) feet elevation; flowering May to November, mostly May to July.

Although browsed to some extent, whitethorn has little forage value, and has become a range pest in some areas because of its encroachment on grasslands, as in Cochise County.

POISONOUS PROPERTIES — Whitethorn, like Johnsongrass, produces hydrocyanic acid under certain conditions.

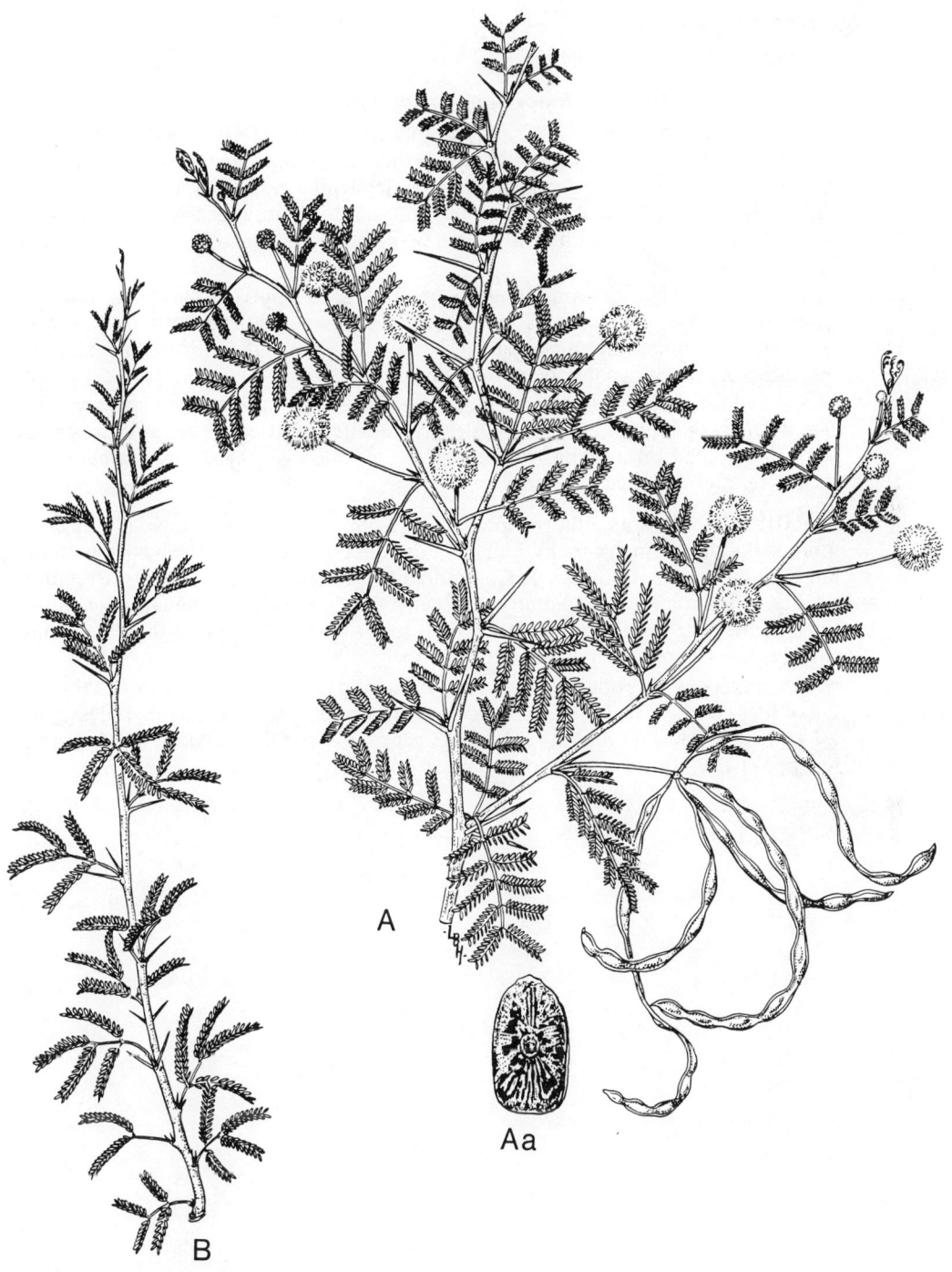

Fig. 75. Whitethorn *(Acacia constricta)*. *A.* Large branch with heads of flowers and curved pods. *Aa.* Seed, enlarged. *B.* Chihuahua whitethorn *(A. constricta* var. *vernicosa)*. Smaller nonflowering branch.

# CAMELTHORN

PEA FAMILY — Leguminosae

### CAMELTHORN — *Alhagi camelorum* Fisch.

DESCRIPTION — A spiny, intricately branched, completely hairless shrub 1½ to 4 feet high, reproducing by seeds, but principally from deep vertical roots and extensive rhizomes. The greenish stems bear slender vicious spines, green with yellow tips, ¼ to 1¾ inches long. The alternate wedgeshaped leaves, yellowish above, bluish green beneath, are ¼ to 1¼ inches long, ⅛ to ½ inch broad, and have very short stalks.

The small pealike flowers, about ⅜ inch long, are pinkish purple to maroon. These occur on short slender spinetipped branches which arise uniformly and in large numbers along the upper part of the stems. When the pods mature and fall off, these branches become persistent spines.

The reddish brown jointed seedpods are curved upward, and commonly have 1 to 4 seeds, or up to 9. The pod is deeply indented, and each seed is clearly outlined like a bead on a string. The kidneyshaped seed is grayish brown, about ⅛ inch long, and 1/12 inch broad.

DISTRIBUTION — Camelthorn, introduced from Asia, grows principally in deep moist soil, but also in dry rocky soil. Abundant in colonies along the banks, bottomlands, and drainage of the Little Colorado and Salt rivers, along canals, irrigation ditches, and sometimes spreading to adjacent cultivated fields; in Navajo, Coconino, Gila, Maricopa, and Yuma counties; 100 to 5,000 feet elevation; flowering May to July, the seedpods persisting until October or November.

The underground roots and rhizomes branch greatly, but usually after they are 2 to 4 feet deep. Once established, a colony increases in size each year. In less than 20 years, the infestation along the canals near Gillespie Dam (Maricopa County) has become continuous for more than 15 miles.

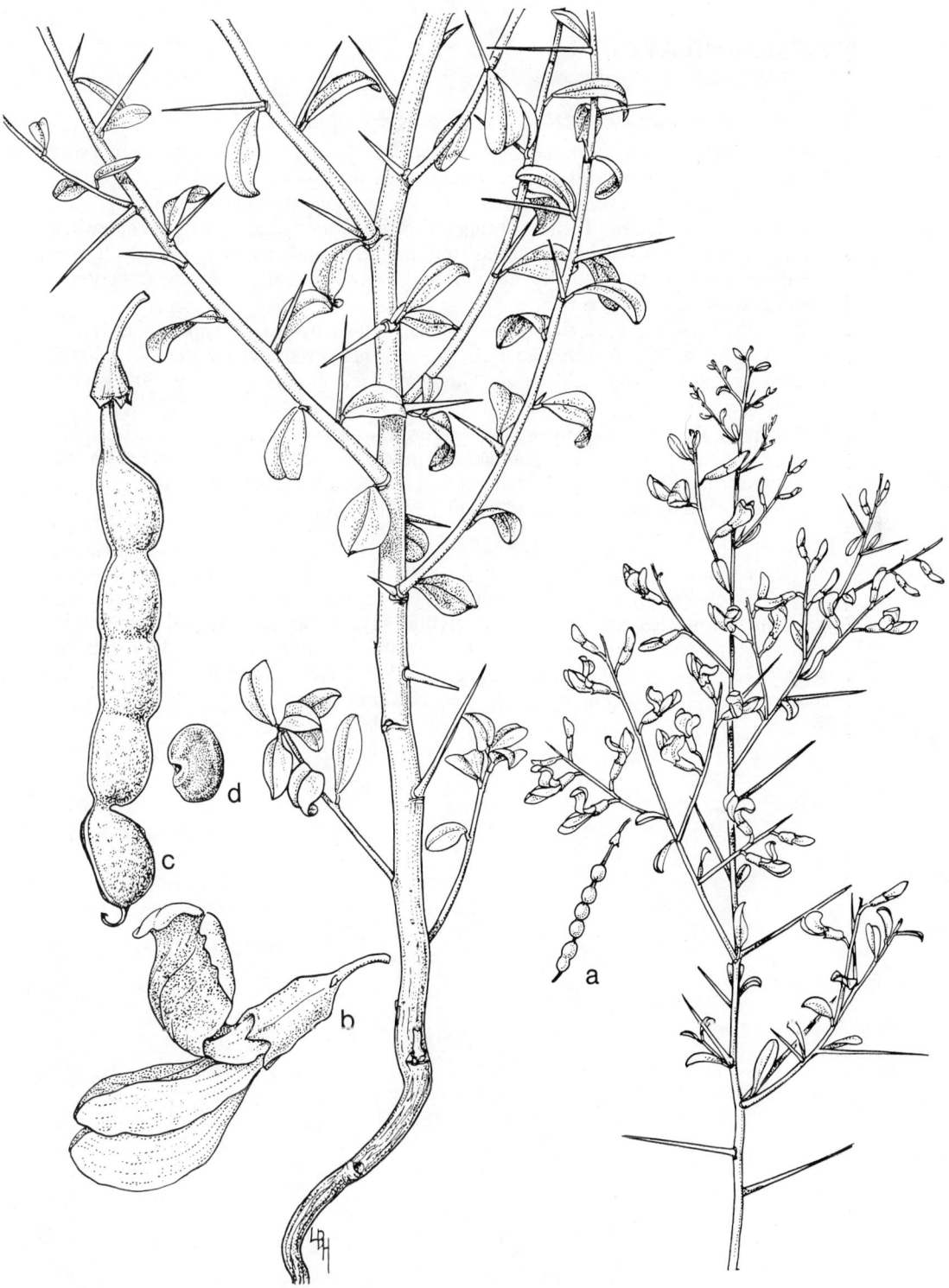

Fig. 76. Camelthorn *(Alhagi camelorum)*. Lower part of plant showing leaves and spines. Also, a flowering branch. *a.* Pod natural size. *b.* Enlarged flower. *c.* Enlarged pod. *d.* Seed.

161

# NUTTALL MILKVETCH
PEA FAMILY — Leguminosae

## NUTTALL MILKVETCH — *Astragalus nuttallianus* DC.

DESCRIPTION — A grayish slender annual, thinly covered with straight grayish hairs lying flat on the surface, reproducing only from seeds. The few weak stems are almost erect, and usually less than a foot high when growing with other plants. On barren soil they are prostrate, branch profusely, and radiate from a somewhat woody taproot, ½ to 3½ feet in diameter, and often appear perennial. The leaves are divided into 3 to 6 (or 8) pairs of small narrow leaflets, the tips rounded, pointed, or with a slight notch.

The small pealike flowers are light purple, fading to nearly white, then drying blue. Two to 6 flowers (and later the pods) are clustered near the end of stiff threadlike stalks, 1 to 3 inches long. The stalks are solitary at the end of the stems and in the leaf axils.

The narrow flat pods, ½ to ¾ inch long and 1/12 to ⅛ inch long, curve upward, are hairy, or may be hairless, and slender-pointed at the top. They persist on the plant until they turn black. The tan colored flattened seeds are narrowly squarish, with a deep notch in one edge, and about 1/12 inch long.

DISTRIBUTION — Nuttall milkvetch is a native plant, and a common lawn pest, particularly in new lawns, yards, ditches, and fields. It is Arizona's most common annual *Astragalus,* and is abundant throughout the state on barren dry or disturbed soil along roadsides, waste places, river bottoms, mesas, slopes, and canyons in southern and northern desert, desert grassland, chaparral, and oak woodland ranges; 100 to 4,000 feet and infrequent to 7,000 feet elevation; flowering February to May, and sometimes again after the summer rains.

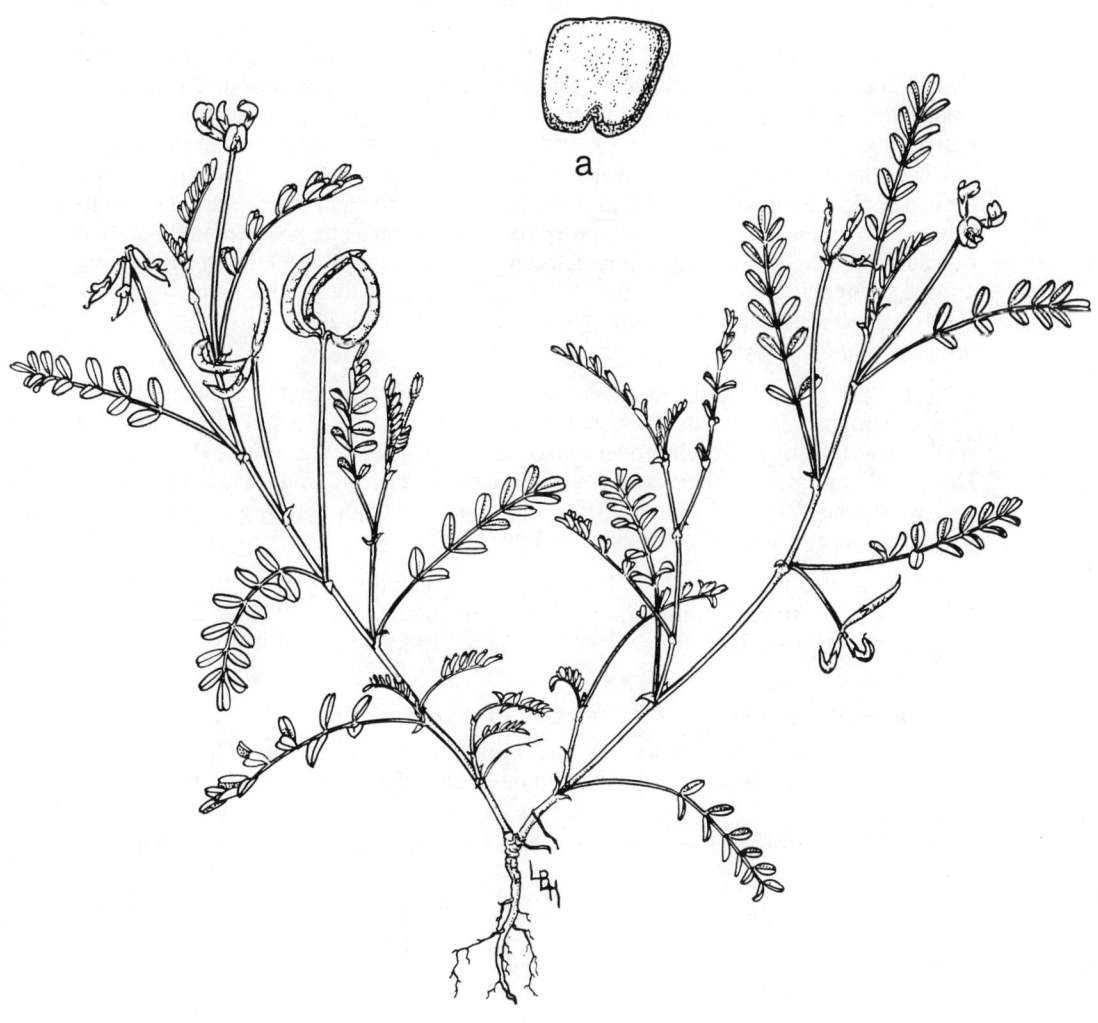

Fig. 77. Nuttall milkvetch *(Astragalus nuttallianus).* Plant with
flowers and pods, the compound leaves with 3 to 6 pairs of leaflets. *a.* Seed.

# WOOTON LOCO, western loco, locoweed, rattleweed
PEA FAMILY — Leguminosae

## WOOTON LOCO — *Astragalus wootonii* Sheldon

DESCRIPTION — A rank, spreading annual, biennial, or shortlived perennial from a thick taproot, reproducing by seeds. The weak stems are branched from the base, erect and bushy at first, but sprawling on the ground in age, usually not over 1 foot long, and mostly less. The leaves are divided into 4 to 11 pairs of small narrow grayish leaflets, 1/12 to ⅛ inch broad, with inconspicuous straight hairs lying on the surface.

The pealike flowers are reddish purple, fading paler, and only ¼ to ⅓ inch long. The flower stalks, 1 to 2½ inches long, arise in the leaf axils, bear about 5 to 10 flowers, and usually do not extend beyond the leaves. The yellowish thin walled pods are papery, and greatly inflated when mature. They are nearly straight along 1 edge, strongly curved on the other, ⅝ to 1⅛ inches long. The reddish brown, flattened seeds are broadly kidneyshaped, about ⅛ inch long, and have a notch on 1 edge.

DISTRIBUTION — Wooton loco is a very common native plant, growing mostly on dry sandy soil along roadsides, mesas, plains, slopes, and washes throughout the state. In wet springs it often covers large areas (e.g. between Window Rock and Holbrook), in both northern and southern desert, desert grassland, and sometimes oak woodland ranges; from 1,300 to 7,000 feet elevation; flowering March to May or June, and sometimes again in August.

POISONOUS PROPERTIES — Wooton loco is definitely known to cause loco poisoning in livestock. Animals do not like it, and usually will not eat it except when hungry for green feed, or after they are addicted to locoweed. It is poisonous even in the dry state.

## HALFMOON LOCO — *Astragalus allochrous* Gray

DESCRIPTION — Halfmoon loco is also known to be poisonous. It is often confused with Wooton loco, and has the same geographical and habitat distribution. The stems are 1 to 2 feet long, the leaves have 11 to 19 pairs of leaflets, 1/12 to ⅓ inch broad, and the pods are larger and more conspicuous, 1⅛ to 1¾ inches long.

Fig. 78. Wooton loco *(Astragalus wootonii)*. Plant with flowers and pods, the compound leaves with 4 to 11 pairs of small leaflets. *a*. Seed.

## DESERT SENNA, cove senna
PEA FAMILY — Leguminosae

### DESERT SENNA — *Cassia covesii* Gray

DESCRIPTION — A bushy perennial covered with fine white hairs reproducing by seeds only. The stems are 1 to 2 feet high, branching from a woody base. The grayish green leaves are 1 to 2 inches long, including the stalks. They are divided into 2 to 4 pairs of large oblong, point tipped leaflets, ⅓ to 1¼ inches long.

The flowers are in several to many stalked clusters at the top of the stems, and in the leaf axils. They have 5 large yellow petals about ½ inch long, with reddish veins, and 10 stamens with large orange anthers. The slightly curved pods, ¾ to 1⅓ inches long, are tipped by a stiff point about ⅛ inch long. They pop open with force when mature, throwing the seeds some distance from the plant. The pinkish brown seeds are pearshaped, deeply wrinkled, and flattened, about ⅛ inch long.

DISTRIBUTION — Desert senna is a very common native weed of dry disturbed soil throughout the state, along roadsides and waste places; also common on rocky slopes, mesas, sandy river bottoms, and washes in desert, northern desert, and desert grassland ranges; mostly from 1,000 to 3,500 feet elevation; flowering from spring until fall, March to October.

### TWOLEAF DESERT SENNA — *Cassia bauhinioides* Gray

DESCRIPTION — Looks almost exactly like desert senna, except the leaves are divided into just 2 leaflets (1 pair), and there are only 2 flowers in each stalked axillary cluster, with no flowers borne at the tip of the stems.

DISTRIBUTION — It has the same distribution and habitat as desert senna, but at slightly higher altitudes, and is more common on northern Arizona desert rangelands; 2,000 to 5,500 feet elevation; flowering from April to September.

Fig. 79. Desert senna *(Cassia covesii)*. Flowering branch with young pods; the leaves are compounded with 2 to 4 leaflets. *a.* Enlarged flower. *b.* seed.

# SLIMPOD SENNA, longpod senna

PEA FAMILY — Leguminosae

## SLIMPOD SENNA — *Cassia leptocarpa* Benth. var. *glaberrima* Jones

DESCRIPTION — A rank smelling bushy very leafy perennial, 2 to 3 feet high, reproducing by seeds. The stems are greatly branched above, often spreading 2 to 4 feet in diameter. The alternate leaves are a deep bright green 3 to 7 inches long, and divided into 4 to 8 pairs of sharp pointed lanceshaped leaflets. These are 1 to 2¾ inches long and nearly hairless, except for the short eyelashlike hairs around the edges. There is a large gland near the base of the leaf stalk.

The showy orange yellow flowers are on long flowering branches at the top of the stems. There are 5 petals, ½ to ¾ inch long, and 10 stamens with large orange anthers. The long slender pods are several in a drooping cluster. They are dull brown, 5 to 10 inches long when mature, and have 1 row of seeds. The seeds are grayish brown, broadly cubical in shape, and about ⅛ inch in diameter.

DISTRIBUTION — Slimpod senna is a native range weed found in very much the same type of places as the other sennas. It is common along roadsides, streams, washes, canyons, and eroded areas in southeastern Arizona desert, desert grassland, and desert shrub ranges, largely in Graham, Pinal, Cochise, Santa Cruz, and Pima counties; 2,500 to 5,500 feet elevation; flowering June to September.

Fig. 80. Slimpod senna *(Cassia leptocarpa* var. *glaberrima).* Flowering branch with immature pods. *a.* Compound leaf with six pairs of leaflets and a gland at the base of the stalk. *b.* Two views of seeds. *c.* Pods.

# HOG POTATO, pignut
PEA FAMILY — Leguminosae

## HOG POTATO — *Hoffmanseggia densiflora* Benth.

DESCRIPTION — A low weak slender stemmed perennial ½ to 1 foot high, which reproduces from seeds, from creeping underground horizontal roots, and from deep seated nutlike tuberous enlargements of the roots. The twice divided leaves are alternate, 2 to 5 inches long including the stalks, and are covered with glandular dots and fine incurved hairs. They are divided into 3 to 5 pairs of primary leaflets (or divisions), each of which is further divided into 5 to 10 pairs of oblong secondary leaflets only ⅛ to ¼ inch long.

The flowers have 5 yellow or orange red petals about ½ inch long, each narrowed into a stalk covered with small glands. The flowers occur along the upper part of the stems, the flowering part only about 2 to 6 inches long and covered with small sticky tackshaped reddish glands. The flattened, slightly curved pods, ⅔ to 1⅔ inches long, are dark reddish brown, and have few to several seeds. The grayish seeds are smooth, flattened eggshaped, and slightly more than ⅛ inch long.

DISTRIBUTION — Hog potato is a very common native weed, often forming large colonies in heavy alkaline soil along roadsides, ditch banks, and waste places, becoming a troublesome pest when it spreads to adjacent cultivated lands and pastures. It is common throughout most of the state, especially in irrigated areas at the lower elevations, 100 to 5,000 feet elevation; flowering April to October.

The tubers are 1½ feet deep or more, and relished by hogs; the Indians considered them quite a delicacy after they were roasted. However, the tubers and the roots together enable this plant to become quickly established, and it should be eradicated from cultivated fields and gardens when it first appears.

Fig. 81. Hog potato *(Hoffmanseggia densiflora)*. Flowering plant covered with small glands, the bicompound leaves with 3 to 5 pairs of primary leaflets, each of which has 5 to 10 pairs of secondary leaflets. *a.* Pod. *b.* Seed. *c.* Tuberous enlargement of the root.

# KINGS LUPINE

PEA FAMILY — Leguminosae

### KINGS LUPINE — *Lupinus kingii* Wats.

DESCRIPTION — A low bushy hairy annual 3 to 8 inches high, with long and fine dense hairs. The leaves are alternate, and divided into 5 (sometimes 6 to 9) leaflets which arise from a common point at the end of the short (½ to 1½ inches long), hairy leaf stalks.

The small pealike flowers, about ⅓ to ½ inch long, are blue or violet, with whitish centers. They are crowded very close together on short flower branches only ½ to ¾ or 1 inch long, which rarely extend beyond the leaves. The mature pods are eggshaped, ⅓ to scarcely ½ inch long, and contain just 2 seeds. The grayish tan seeds are rounded, plumpish and scarcely ⅛ inch in diameter, with the surfaces smooth.

DISTRIBUTION — A very common native weed in rocky clay or disturbed soil along highways, old fields, vacant lots, and waste places. Also locally abundant on eroded or overgrazed meadows and openings in yellow pine, and sometimes in pinyon-juniper ranges, mostly in northern Arizona, from Apache to Coconino and Yavapai counties, rare southward in Graham, Cochise, and Pima counties; 5,500 to 8,000 feet elevation; flowering May to September, mostly June to August.

Kings lupine contains poisonous alkaloids, but is not known to be the cause of livestock poisoning.

### LOW LUPINE — *Lupinus pusillus* Pursh

DESCRIPTION — A low annual similar in appearance to kings lupine. Differing in that the flower branches are 1 to 2½ inches long, with the flowers not as crowded; also the pods are oblong, ½ to ¾ inch long, and are indented or constricted between the 2 seeds. Low lupine occurs in dry sandy soil of mesas, canyons, and wastelands in northern Arizona, mostly on pinyon-juniper ranges from Apache to Mohave counties; 4,500 to 8,000 feet elevation; flowering May to June.

POISONOUS PROPERTIES OF NATIVE LUPINES — Among the 22 kinds of native lupines in Arizona, 2 are important elsewhere as the serious cause of livestock poisoning, chiefly to sheep. Silvery lupine (*L. argenteus* Pursh), the most common perennial lupine in the forests of northern Arizona at 7,000 to 10,000 feet elevation, is very serious in most western states, and low lupine (*L. pusillus*) in Kansas. Lupines may be expected to cause some sheep fatalities in Arizona.

172

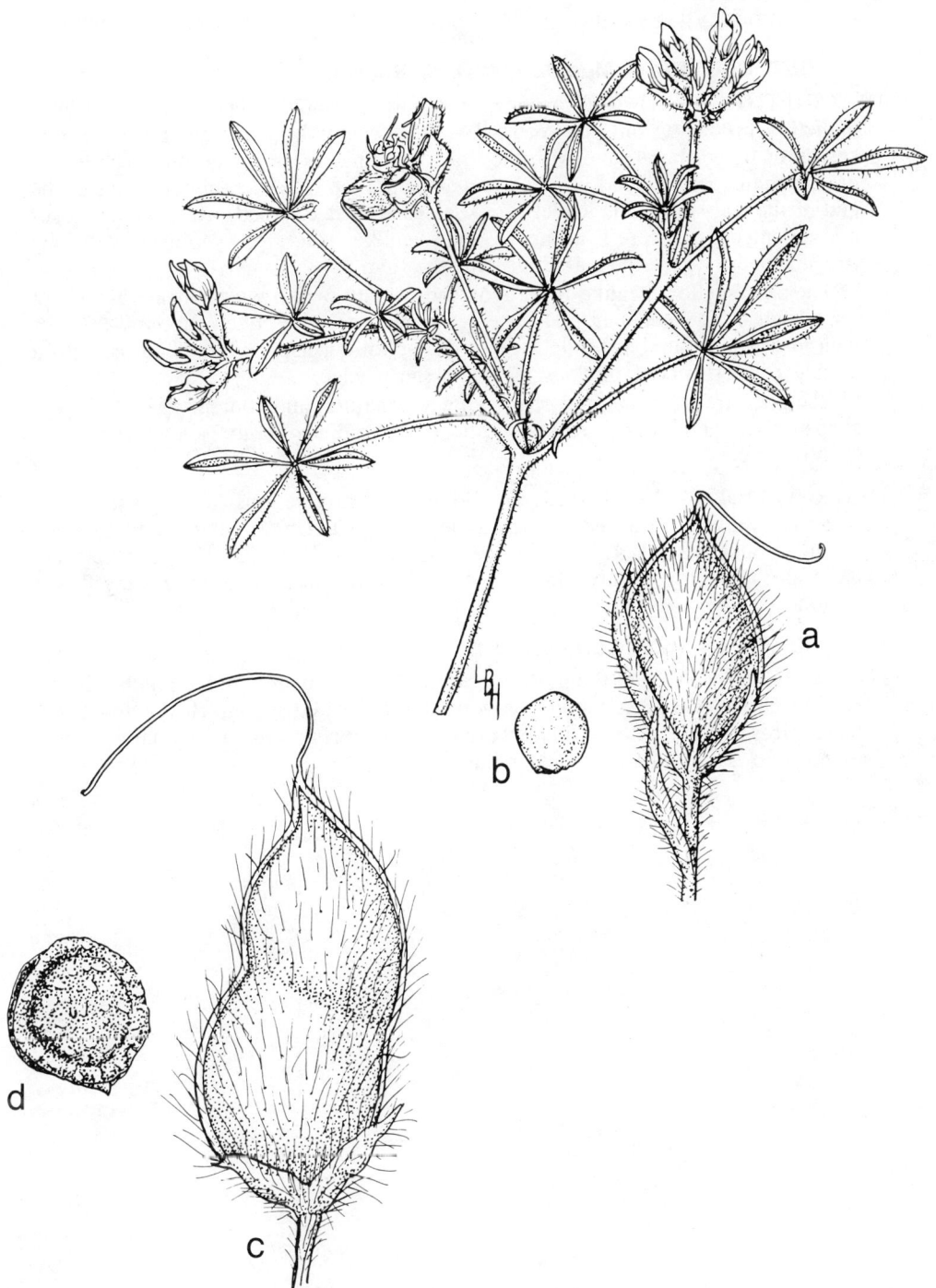

Fig. 82. Kings lupine *(Lupinus kingii)*. Flowering branch with
young pods, the palmately compound leaves with 5 (6 to 9) leaflets.
*a.* Pod containing two seeds only. *b.* Seed. *c.* Low lupine *(L. pusillus).*
Pod, also with two seeds but constricted between the seeds. *d.* Seed.

173

# BUR CLOVER

PEA FAMILY — Leguminosae

## BUR CLOVER — *Medicago hispida* Baertn.

DESCRIPTION — A bright green nearly hairless annual, or sometimes winter annual, reproducing only by seeds. The weak stems branch from the base, and spread or lie on the ground with the tips ascending, ¼ to 2 feet long. The leaves are alternate, and divided into 3 leaflets which arise from a common point at the end of the leaf stalk. The leaflets are somewhat wedgeshaped, with toothed edges and indented tip. There is a pair of small leaflike structures (stipules) with long irregular teeth where the leaf stalk joins the stem.

The small yellow pealike flowers are borne 3 to 5 in a cluster near the end of short stalks. The peculiar pods are net-veined, spirally coiled, about ¼ to ⅓ inch in diameter, and are indehiscent. The spiral consists of 2 or 3 turns, with a double row of curved prickles along its sharp edge.

The pods are straw colored or brown when mature, and contain several kidney-shaped seeds. The seeds are yellowish or tan colored, about ⅛ inch long, and slightly sticky.

DISTRIBUTION — Bur clover is an Old World introduction. Although it is high in forage value, it is a very troublesome weed in Bermuda lawns, and to some extend in alfalfa seed crops in southern and central Arizona; 100 to 5,500 feet elevation; flowering mostly March to May, but in moist situations it may flower at almost all seasons.

## ALFALFA — *Medicago sativa* L.

DESCRIPTION — One of the most important of all cultivated forage plants, alfalfa, also an Old World introduction, is very closely related to bur clover and black medic. It is entirely different in appearance, since it is an erect perennial with blue or violet flowers, but its seeds are similar.

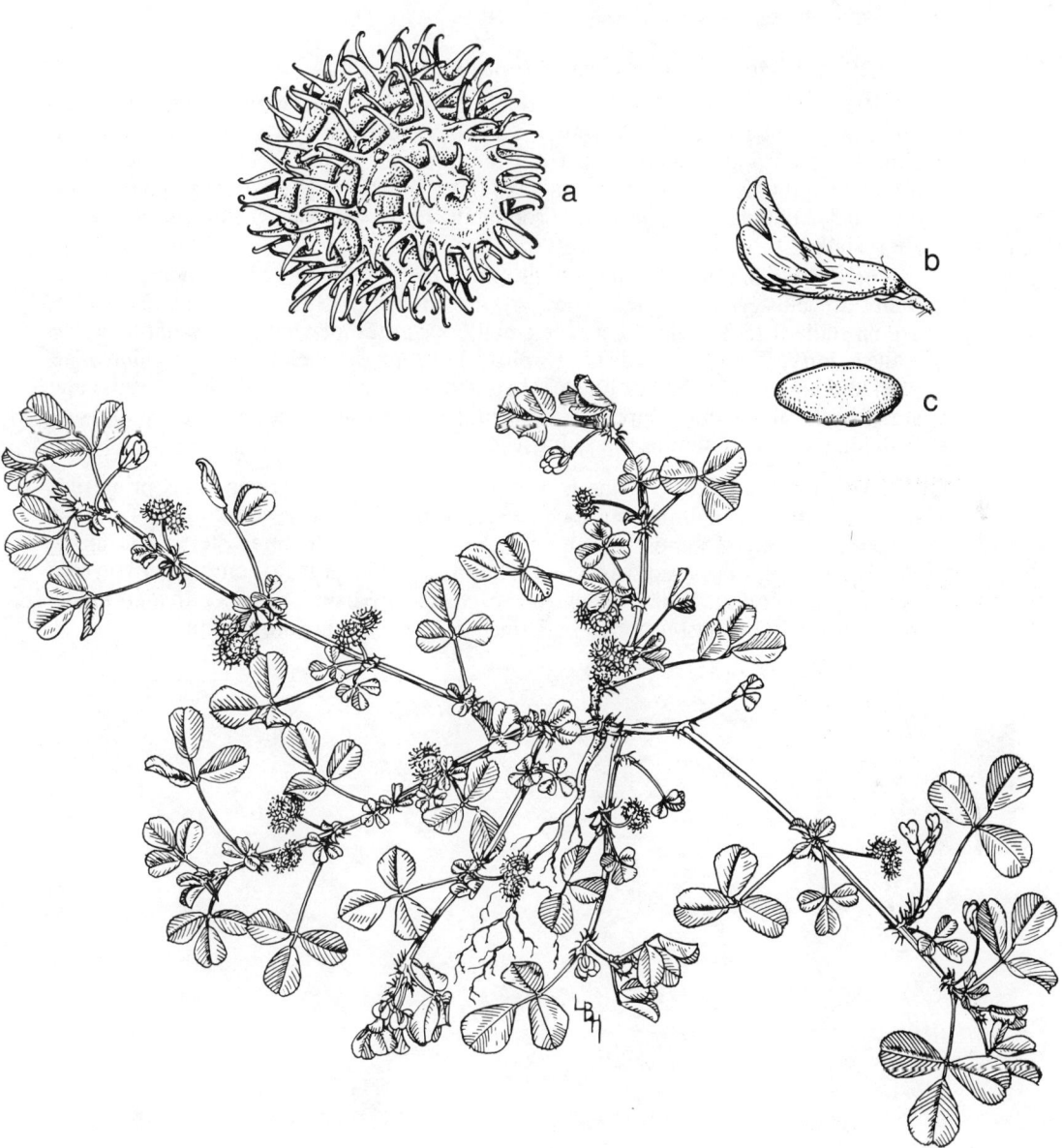

Fig. 83. Bur clover *(Medicago hispida).* Plant with flowers and pods; the leaves are trifoliolate (i.e., have three leaflets). *a.* Pod, spirally coiled, with prickles. *b.* Flower, pealike. *c.* Seed.

# BLACK MEDIC, nonesuch, hop medic
PEA FAMILY — Leguminosae

## BLACK MEDIC — *Medicago lupulina* L.

DESCRIPTION — A low trailing annual, biennial, or perennial reproducing by seeds only. The slender hairy 4-angled stems are prostrate or partly so, branching from the base and radiating out from the taproot, ½ to 2 feet long. The leaves are divided into 3 leaflets, very similar to those of bur clover, with the tips rounded and finely toothed, and the veins prominent. However, the central leaflet is on a short stalk.

The flowers are also pealike, small and bright yellow, about ⅛ inch long, but many are clustered together. They are crowded into short dense spikes, which are on stalks 1 to 3 inches long. The small pods are somewhat kidneyshaped, thick walled, curved, and with distinct veins, 1/12 to ⅛ inch long. They have no prickles, are usually hairy, black when mature, and contain only 1 seed. The small seeds, also kidneyshaped, are greenish yellow to brown, about 1/16 to 1/12 inch long, and very similar to alfalfa seed.

DISTRIBUTION — Black medic is a native of Eurasia. In Arizona it is primarily, as is bur clover, a nuisance in lawns where it thrives, and in alfalfa seed fields. It is found as a weed almost throughout the state in moist, often sterile soil along roadsides, in gardens, and waste places; up to 8,000 feet elevation; flowering from April to September. Where black medic grows in pastures or uncultivated fields, it can scarcely be classed as a weed since it is so high in forage value.

Fig. 84. Black medic *(Medicago lupulina)*. Semi-prostrate plant with flowers and pods on stalked, short spikes, the leaves with three leaflets. *a.* Pealike flower. *b.* Pod.

177

# WHITE SWEETCLOVER
## PEA FAMILY — Leguminosae

### WHITE SWEETCOVER — *Melilotus albus* Desr.

DESCRIPTION — A tall erect hairless annual or biennial, rarely perennial, which reproduces only by seeds. The stem is erect, branching above 1 to 6 feet high. The leaflets are similar to those of annual yellow sweetclover. The flowers are also similar, but are white instead of yellow, and larger, about ⅕ inch long. They are likewise arranged in long narrow spikelike clusters along the upper half of the flower stalks, but the stalks are 2 to 8 inches long. The pods are slightly larger, 1/12 to slightly more than ⅛ inch long. The seed is oblong to oval, about 1/12 inch long, notched near one end, yellowish green to brown, with a smooth, not roughened surface.

DISTRIBUTION— White sweetclover was introduced from Eurasia. As a weed it is not as serious as annual yellow sweetclover. A pest in cotton, alfalfa, grapes, and in field ends and borders of cultivated fields; also in other cultivated crops. It is a common weed in moist sandy soil, scattered throughout most of the state along roadsides, ditches, fences, and creeks; 100 to 7,500 feet elevation; flowering March until October.

Both white sweetclover and annual yellow sweetclover are excellent honey plants. The herbage is very fragrant when dried, but has a peculiar odor when fresh. They rate high in forage value, and animals soon acquire a taste for them.

### YELLOW SWEETCLOVER — *Melilotus officinalis* (L.) Lam.

DESCRIPTION — Very similar to white sweetclover, but the flowers are yellow and the stems are less erect and not as stout. Not as common nor as widespread in Arizona as annual yellow sweetclover or white sweetclover.

Fig. 85. White sweetclover *(Melilotus albus)*. Flowering branch, the pinnately compound leaves have three leaflets. *a*. Seed. *b*. Flower stalk with spikelike inflorescence.

## ANNUAL YELLOW SWEETCLOVER, sour clover

PEA FAMILY — Leguminosae

### ANNUAL YELLOW SWEETCLOVER — *Melilotus indicus* (L.) All.

DESCRIPTION — Annual or sometimes a biennial which reproduces only by seeds. The main stem is erect, with many spreading branches from above the base, 1½ to 3 feet high. The leaves are alternate, and divided into 3 leaflets as in bur clover, but are much larger. They are ½ to 1¼ inches long, toothed above the middle with the tip indented or blunt, and a reddish bar along the midrib.

The yellow flowers are pealike, and less than ⅛ inch long. They are numerous and spikelike along the upper half of slender flower stalks, which are 2 to 4 inches long and arise from the upper leaf axils. The flower buds are erect, but soon bend downward. The small globeshaped pods are swollen on one edge, wrinkled, about 1/12 inch long, and contain 1 or 2 seeds. The seed is eggshaped, about 1/16 inch long, dark greenish brown, with a honeycombed and roughened surface.

DISTRIBUTION — Annual yellow sweetclover, introduced from Eurasia, is often a pest in lawns. It is found throughout the state in damp soil on bottomlands, gardens, roadsides, fields, ditches, and waste places. Often this weed is a serious problem in croplands, as small grains, citrus orchards, and winter vegetables; frequently a nuisance in alfalfa hay, but not often in alfalfa seed; 100 to 7,500 feet elevation; flowering November to June.

Fig. 86. Annual yellow sweetclover *(Melilotus indicus)*. Erect plant
with spikelike inflorescences and pendulous flowers; the leaves are compound
and have three leaflets. *a.* Seedling. *b.* Pealike flower. *c.* Seed.

181

## LAMBERT CRAZYWEED, Lambert loco, point loco, purple loco, white loco
PEA FAMILY — Leguminosae

### LAMBERT CRAZYWEED — *Oxytropis lambertii* Pursh var. *bigelovii* Gray

DESCRIPTION — A stemless hairy perennial from a woody taproot system with a branching base, which reproduces only by seeds. There are no true stems, and all the leaves come from the base of the plant. These leaves are 3 to 10 inches long, and divided once into 7 to 9 pairs of silky hairy leaflets ½ to 1 inch long.

The many flower stalks are 4 to 16 inches long, nearly naked ⅔ of their length, and bear numerous flowers on the upper part. The showy flowers are pealike, commonly purple or sometimes white, ½ to 1 inch long. The oblong woody seedpods are hairy, ¾ to 1¼ inches long, including the slender beak about ¼ inch long. The reddish brown, somewhat kidneyshaped seeds are angular and flattened, about 1/12 inch long.

DISTRIBUTION — Lambert crazyweed is a native weed, growing in sandy or open slopes and plains in the yellow pine forests. It is very common and abundant in some areas where it has increased and spread on overgrazed and "sheeped out" areas in the open pine ranges of eastern Arizona, from Apache to Coconino counties, and in the mountains of Greenlee, Graham, and Cochise counties; 5,000 to 8,000 feet elevation; flowering June to September.

POISONOUS PROPERTIES — Lambert crazyweed and a few species of *Astragulus* (locoweeds) cause "locoism" in horses, cattle, and sheep. Poisoning in Arizona occurs principally among cattle in the spring. Losses have been reported from Navajo county, near Showlow, Snowflake, and Holbrook; also in the Pinaleno Mountains, Graham County.

Blue loco (*Astragalus lentiginosus* Dougl. var. *diphysus* [Gray] Jones) might also be involved, since it is known to be poisonous, and has about the same distribution as Lambert crazyweed. Cattle losses caused by locoweed are also reported from the Canelo Hills, Sonoita, Patagonia, and Elgin areas in Santa Cruz county. These are probably due to sheep loco *(Astragalus nothoxys* Gray), since Lambert crazyweed does not occur there. For a full discussion of locoweed and crazyweed poisoning in Arizona, see *Livestock-Poisoning Plants,* by Schmutz, Freeman, and Reed.

Fig. 87.  Lambert crazyweed *(Oxytropis lambertii)*.  Flowering plant,
the compound leaves with 7 to 9 pairs of silky hairy leaflets. *a.* Pod.  *b.* Seed.

# WESTERN HONEY MESQUITE and VELVET MESQUITE

PEA FAMILY — Leguminosae

## WESTERN HONEY MESQUITE AND VELVET MESQUITE —
### *Prosopis juliflora* (Swartz.) DC. var. *torreyana* Benson and var. *velutina* (Woot.) Sarg.

DESCRIPTION — A spiny deciduous shrub or small tree up to 30 or exceptionally 55 feet high, with a trunk 1 to 4 feet in diameter. Usually armed with stout yellowish, nearly straight spines arising in pairs, ¼ to 3 inches long. The leaves, 3 to 8 inches long, are first divided into 1 to 2 pairs of primary divisions. Each of these is again divided into about 10 to 28 pairs of finely hairy or hairless secondary leaflets, ⅛ to ¾ inch long.

The small fragrant greenish yellow flowers are crowded on stalked spikes 2 to 5 inches long. The flat tan colored leatherish pods are finely hairy or hairless, 3 to 8 inches long, with a sweetish pulp. The rough bark separates into dark strips, and the wood is hard, reddish brown, with thin yellow sapwood.

DISTRIBUTION — Mesquite is abundant throughout southern and central Arizona, and also occurs in northern Arizona in Coconino and Mohave counties, 1,000 to 5,000 (rarely 6,000) feet elevation; flowering March to August, principally May to June. It is a common tree along the watercourses, washes, and alluvial bottoms where ground water is available. In some areas, the roots may penetrate to depths of 60 feet.

Mesquite is abundant, and has become a serious range problem on the mesas and slopes of the deserts and desert grassland ranges, and occasionally lower oak woodlands where it is often a shrub. The carrying capacity of many ranges has been seriously reduced due to its tremendous increase. Dissemination of the seeds in cattle dung has been an important factor in this invasion. Mesquite pods are relished by all livestock, which, unlike most other pea pods, do not shed their seeds.

POISONOUS PROPERTIES — Livestock are sometimes bothered after eating mesquite beans. This is not due to a poison, but to the formation of large hard balls from the long, tough, stringy margins from green or rain-soaked dried pods. Dried beans are not harmful, as the thick fibers easily break into short pieces when eaten.

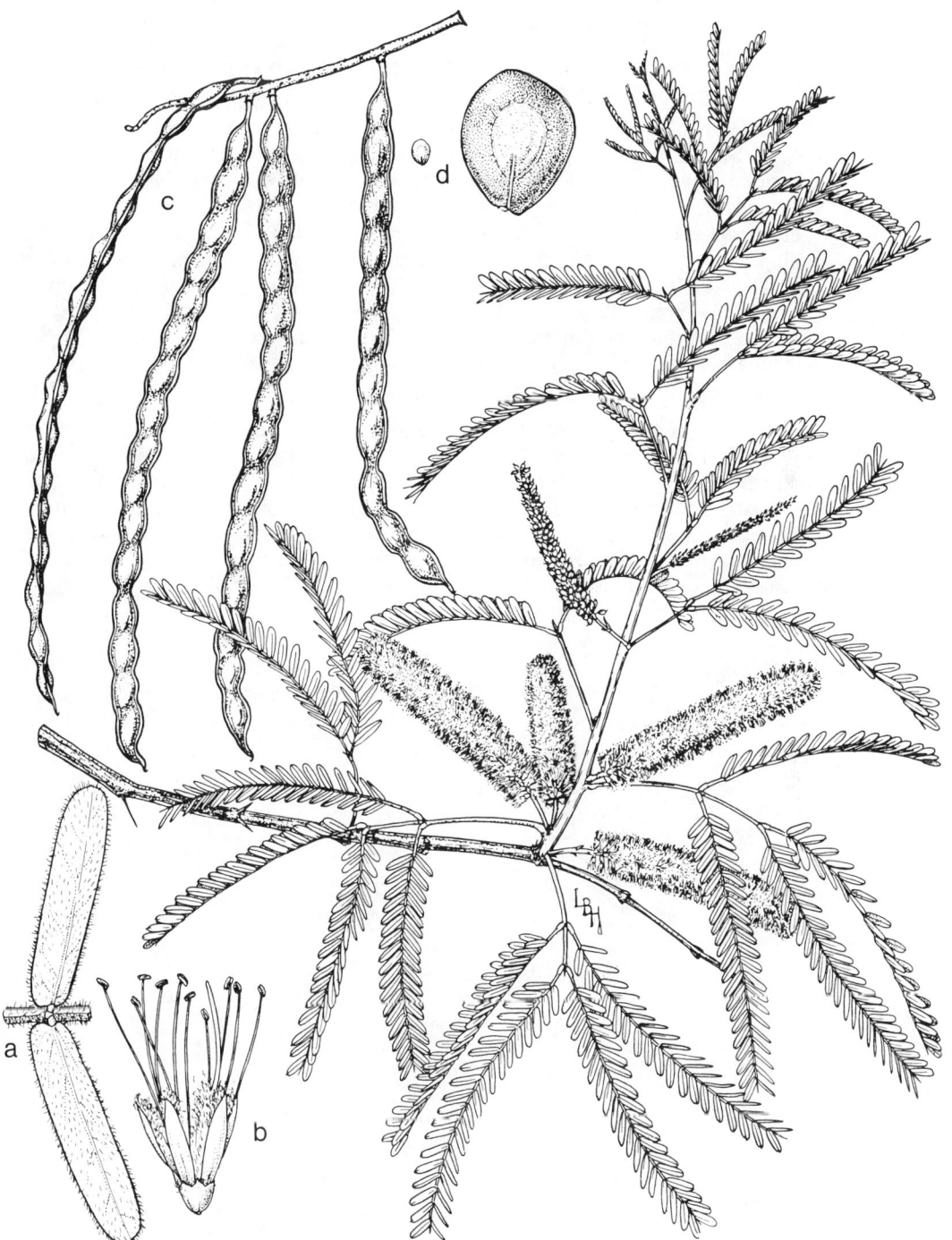

Fig. 88.  Western honey mesquite and velvet mesquite *(Prosopis
juliflora* and vars.).  Branch with flower spikes, the bipinnate leaves with
1 to 2 pairs of primary leaflets and 10 to 28 pairs of secondary leaflets.
*a.* Pair of secondary leaflets. *b.* Flower. *c.* Seed pods. *d.* Seeds, one enlarged.

# HEMP SESBANIA, Colorado River hemp, big-podded sesbania
PEA FAMILY — Leguminosae

## HEMP SESBANIA — *Sesbania exaltata* (Raf.) Rydb.
(*S. macrocarpa* Muhl.)

DESCRIPTION — Tall, entirely hairless annual, 3 to 13 feet high, which reproduces by seed. Smooth, sparingly branched stems with widespreading branches. Large compound leaves, alternate, 2 to 12 inches long, with 10 to 40 pairs of small leaflets about ½ to 1 inch long, and ¼ inch broad.

The pealike flowers are yellow, often purple spotted, about ½ inch long, borne on short flowering stalks in the axils of the upper leaves, with 2 to 6 flowers per stalk. The fruit is a slender pod, 4 to 8 inches long and about ⅛ inch broad, tipped with a slender beak. There are 15 to 40 seeds in each pod, with thick partitions between each seed when mature. The dark brown shiny seeds are capsuleshaped with squarish ends, about ⅛ inch long.

DISTRIBUTION — Hemp sesbania is a native weed growing along the Colorado River bottom lands of southern Mohave and Yuma counties; 100 to about 4,000 feet elevation; flowering March to October. Sometimes grown as a green manure crop in citrus orchards and farms in southern Arizona, then spreading along irrigation ditches and roadsides. After using it as a cover crop in the Yuma region, it appeared in about 70 percent of the cotton fields in that area. Hemp sesbania is a fiber plant, producing very strong smooth threads which are used by the Yuma Indians for nets and fish lines.

Fig. 89.  Hemp sesbania *(Sesbania exaltata)*. Flowering branch
with young pods, the large compound leaves with 10 to 40 pairs of
leaflets. *a.* More mature pod.  *b.* Enlarged seed.

# SILKY SOPHORA

PEA FAMILY — Leguminosae

## SILKY SOPHORA — *Sophora nuttalliana* B. L. Turner
## (*S. sericea* Nutt.)

DESCRIPTION — A low, silky hairy, silver perennial which reproduces by seeds and deep horizontal roots and rhizomes. The weak stems are often woody, and branch at the base with some spreading and others erect, 2 to 12 inches long. The alternate leaves are 1 to 2 inches long, and divided once into 4 to 12 pairs of small leaflets, ¼ to nearly ½ inch long, with fine silky hairs pressed close to the surface.

The white or yellowish flowers are pealike, ½ to ¾ inch long, and occur on short flower branches 2 to 5 inches long at the end of the stems. The pods are somewhat woody, ½ to 1½ inches long, contain 1 to 4 seeds, and are tipped by a sharp pointed beak. The pods are constricted between the seeds, so each seed is clearly outlined. The seeds are kidneyshaped, tan colored with the surface smooth, and ⅛ to ¼ inch long.

DISTRIBUTION — Silky sophora is a native plant, growing in colonies in sandy or heavy disturbed soil, as where floodwaters collect. It is abundant locally, often covering extensive areas, and becoming a weed in cultivated fields, roadsides, and on the open ranges on sandy creek bands, in swales, and bottomlands. Found in northern Arizona from Apache to Mohave and Yavapai counties, also in Cochise and Pima counties; 3,500 to 7,000 feet elevation; flowering April to June. Silky sophora is hard to eradicate because of its underground stems, and should not be allowed to become established.

POISONOUS PROPERTIES — Silky sophora is reported to cause livestock poisoning if eaten in large amounts, and the seeds are known to contain a poisonous alkaloid. Animals seldom eat this plant, and no livestock poisoning due to silky sophora in Arizona is known definitely.

Fig. 90. Silky sophora *(Sophora sericea)*. Flowering plant with horizontal roots, the pinnately compound leaves with 4 to 12 pairs of leaflets. *a.* Pod. *b.* Two views of seed.

# WHITE CLOVER

PEA FAMILY — Leguminosae

## WHITE CLOVER — *Trifolium repens* L.

DESCRIPTION — A hairless perennial with prostrate stems, which reproduces by seeds and creeping stems which root at the joints. The leaves are alternate, and divided into 3 leaflets. The leaf stalks are long and erect, although rising from the prostrate stems.

The flower stalks, also erect from the prostrate stems, are much longer than the leaf stalks, 2 to 9 inches long. The flowers are crowded into globelike heads ½ to 1¼ inches broad, at the end of the flower stalks. The pealike flowers are white or pink tinged, and bend downward in age. The petals do not fall off, but remain on the flower after withering. The small pod contains 3 to 4 seeds. The seeds are globeshaped to kidneyshaped, and about 1/16 inch long.

DISTRIBUTION — White clover was introduced from Europe. It grows in moist soil, and is a pest in lawns throughout the state. It also occurs in moist meadows in the yellow pine and spruce fir ranges, where it can scarcely be classed as a weed since it is high in forage value.

190

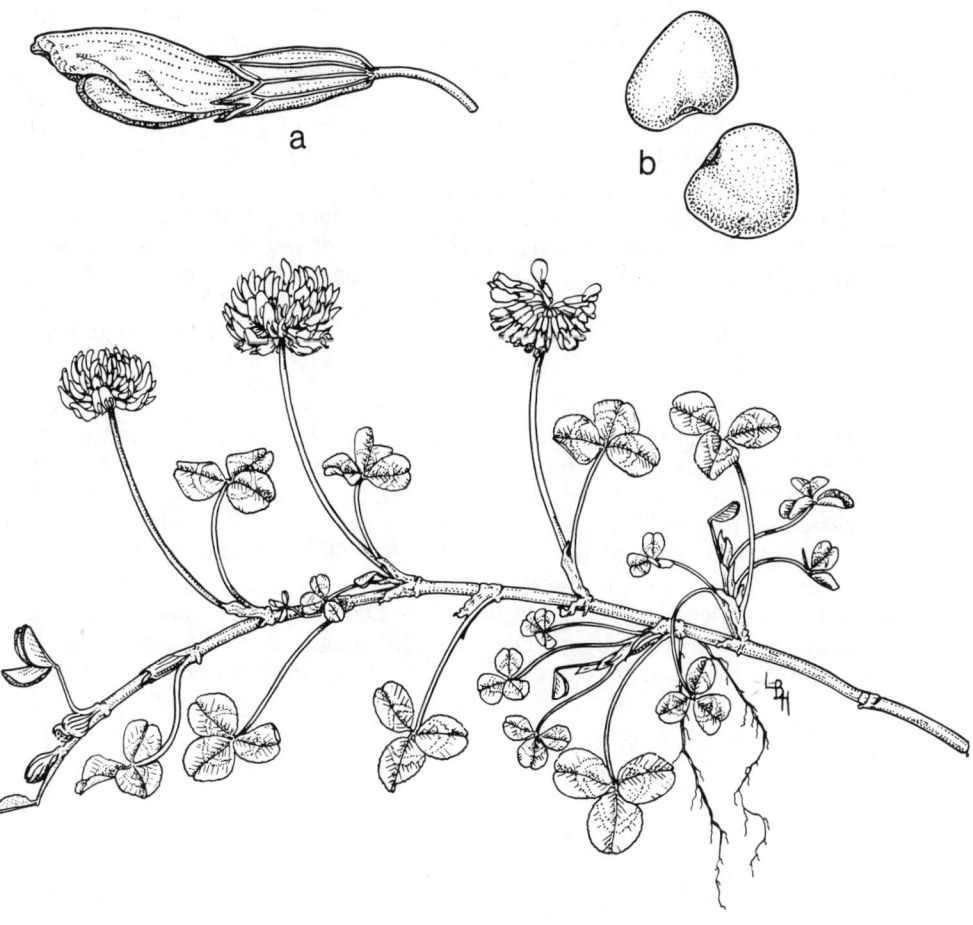

Fig. 91. White clover *(Trifolium repens)*. Prostrate branch rooting at a node, with stalked heads of flowers; the compound leaves are trifoliolate. *a.* Flower. *b.* Two views of a seed.

## REDSTEM FILAREE, filaree, alfilaree, alfilaria
GERANIUM FAMILY — Geraniaceae

### REDSTEM FILAREE — *Erodium cicutarium* (L.) L'Hér.

DESCRIPTION — A dark green annual, winter annual, or biennial which reproduces by seeds only. The many branched stems may be erect, spreading, or prostrate with the tips ascending, ¼ to 2 feet long, often covering areas 2 to 3 feet or more in diameter. The leaves at first form a rosette close to the ground, and are alternate, hairy, and ½ to 4 inches long. They are divided into 3 to 7 or more pairs of stalkless leaflets, which are further divided into many fine segments.

The flowers are in an umbrellalike cluster at the end of long slender stalks arising from the leaf axils. The 5 rose purple petals are ¼ inch or less long, and drop off very quickly. The unusual long needlelike fruits split into 5 one-seeded fruits at maturity. One seed is enclosed in each of these single, spindleshaped, very hard fruits. The fruit is about ⅕ inch long, with a very sharp pointed base, and ends in a slender wiry beak 1 to 1¾ inches long. These "tails" are tightly twisted and corkscrewlike when dry, but uncoil when wet, and can drive the seed into the hardest soil.

DISTRIBUTION — Redstem filaree is native to Europe, and probably was introduced by the Spaniards. It is common in moist soil and a nuisance, particularly in winter lawns, also in flower beds, yards, gardens, fields, and roadsides below 5,000 feet elevation from February to May. Common in the same type of places from April to October at the higher elevations, about 5,000 to 7,500 feet.

It is abundant throughout the state on plains, mesas, and slopes; 100 to 7,000 feet elevation; flowering February to July or to October at elevations to 8,500 feet or higher. Because of its abundance and high forage value, redstem filaree is a very important plant on many Arizona ranges during a short period in the spring.

192

Fig. 92. Redstem filaree *(Erodium cicutarium)*. Branch with flowers and fruits.
*a*. Basal rosette of young plant. *b*. Flower. *c*. Single fruit with one seed
enclosed and corkscrewlike "tail."

193

# CREEPING WOODSORREL, creeping oxalis
WOODSORREL FAMILY — Oxalidaceae

## CREEPING WOODSORREL — *Oxalis corniculata* L.

DESCRIPTION — A spreading to prostrate perennial with weak creeping stems, 3 to 8 inches long, from a slender taproot. Reproducing by seeds, by stems rooting at the joints, and sometimes from slender underground rhizomes. The alternate leaves are divided into mostly 3 broadly heartshaped leaflets borne at the tip of the long leaf stalks, which have sharp tasting juice. The green, purplish, or bronze leaflets, closing and drooping at night, are hairless or thinly hairy.

The flowers have 5 yellow petals, ⅛ to ⅓ inch long. They occur in clusters of 1 to 5 at the end of a slender flower stalk arising from the leaf axils. The yellowish seedpods are erect, but their short stalks are bent sharply downward. They are cylindrical, 5-angled, hairy, ⅓ to 1 inch long, and pointed at the tip. When the many seeds are mature, the seedpods open explosively, throwing the seeds some distance. The tiny reddish brown seeds are somewhat eggshaped but flattened, with 7 to 10 ridges on each face, and about 1/25 inch long.

DISTRIBUTION — Creeping woodsorrel grows in dry or moist, usually shaded soil. Principally a nuisance in lawns and greenhouses, and often found close to buildings, this European introduction can be very aggressive and persistent when it becomes established in a lawn. Common in northern and southern Arizona; 100 to 8,000 feet elevation; flowering February to November.

194

Fig. 93. Creeping woodsorrel *(Oxalis corniculata)*. Prostrate branch
rooting at the joints with flowers and seedpods, the compound leaves trifoliolate.
*a.* Flower. *b.* Seedpod. *c.* Two views of seed.

195

### ORANGE CALTROP, Mexican poppy, summer poppy, Arizona poppy
CALTROP FAMILY — Zygophyllaceae

### ORANGE CALTROP — *Kallstroemia grandiflora* Torr.

DESCRIPTION — An erect, reclining, or prostrate annual, covered with long rough yellowish hairs, reproducing only by seeds. The stiffly hairy stems, branching from the base, are ½ to 2 or more feet long. The leaves are opposite, 1¼ to 3 inches long, and divided into 5 to 7 pairs of smooth margined hairy leaflets.

The large flowers have 5 deep orange petals ⅔ to 1¼ inches long. They are solitary on slender stalks, ½ to 2 inches long in the leaf axils. The greenish, somewhat pearshaped seedpods have cross ridges on the back, and are tipped by a long beak ⅓ to ⅜ inch long. The pods split into 8 to 12 wedgeshaped, one-seeded nutlets at maturity. The nutlets are 3-angled, about ⅛ inch long, with 2 brownish and netted veined faces.

DISTRIBUTION — This beautiful native plant is common and very colorful on sandy or gravelly soil on mesas, slopes, washes, roadsides, and bottom lands in southern and central Arizona, from Greenlee to Yavapai to Yuma counties and southward; 100 to 5,000 feet elevation; flowering February to September, but mostly in July and August. It becomes troublesome when it volunteers in adjacent cotton and other crop fields in the irrigated valleys. In rich soil, as between cotton rows, the prostrate stems may be 4 feet long.

### CALIFORNIA CALTROP — *Kallstroemia californica* (Wats.) Vail

DESCRIPTION — An annual very similar to orange caltrop, but the petals are only ¼ inch. The seedpod beaks are no longer than ⅓ inch, and the backs of the seedpods have sharp tubercles. A native plant growing in the same type of places as orange caltrop, but found in northern as well as southern Arizona; 100 to 7,000 feet elevation; flowering May to October.

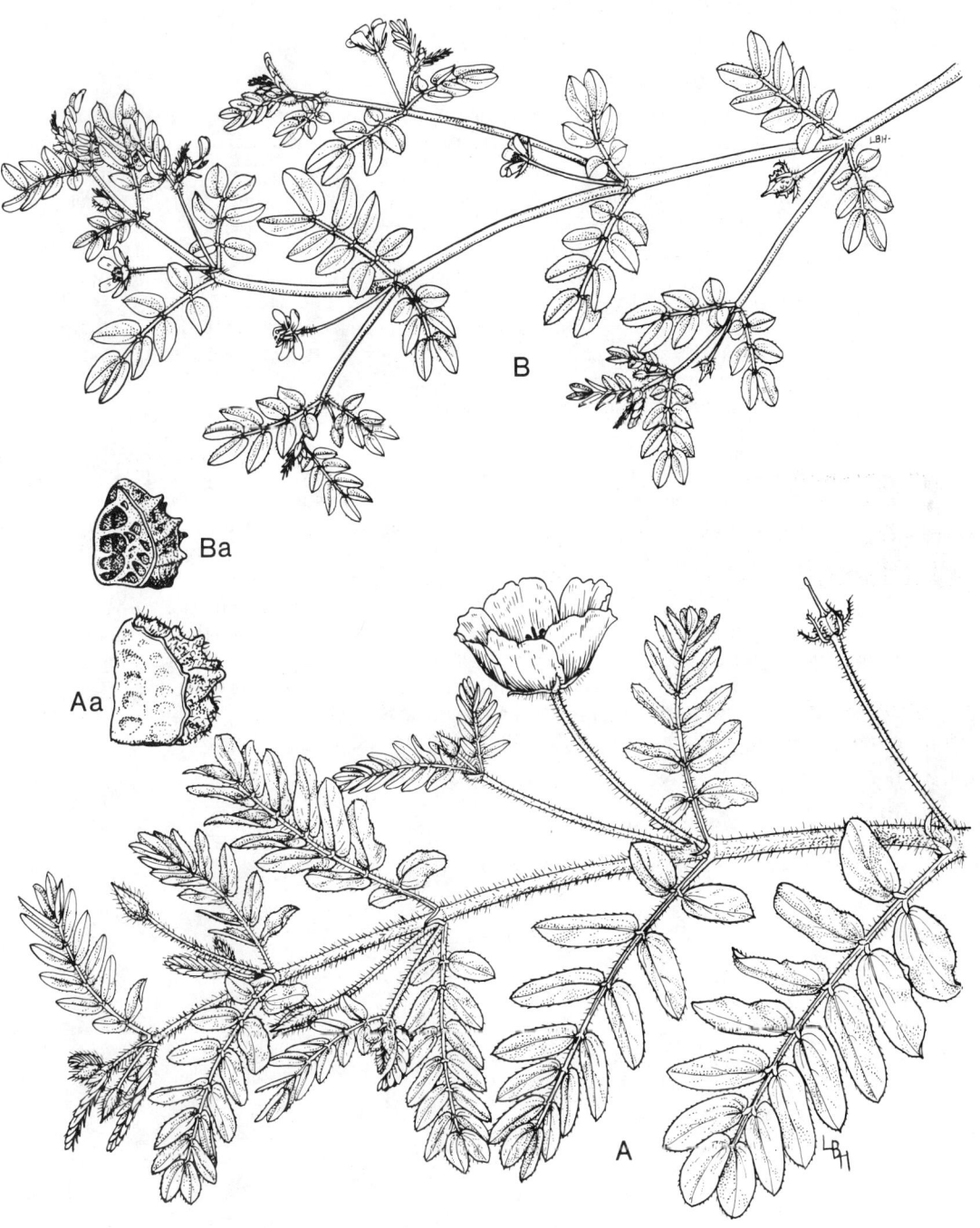

Fig. 94. *A.* Orange caltrop *(Kallstroemia grandiflora).* Flowering branch with beaked seedpod; the compound leaves have 5 to 7 pairs of leaflets. *Aa.* One-seeded nutlet. *B.* California caltrop *(K. californica).* Branch with flowers and seedpod. *Ba.* One-seeded nutlet.

197

# PUNCTUREVINE, bullhead, goathead
CALTROP FAMILY — Zygophyllaceae

## PUNCTUREVINE — *Tribulus terrestris* L.

DESCRIPTION — A restricted noxious weed in Arizona, puncturevine is a prostrate silky hairy annual from a shallow taproot, reproducing only by seeds. The trailing stems, 1 to 5 (or to 8) feet long, branching and radiating from the root, often form dense mats several feet in diameter. Or the stems may be nearly erect where growing in competition for light. The leaves are opposite, divided into 4 to 7 pairs of oblong leaflets ⅛ to ½ inch long.

The bright yellow flowers, with 5 petals which are open only in the mornings, are solitary on short stalks in the axils of the leaves. The seedpod consists of a cluster of 5 flat spiny burs or nutlets, which break apart at maturity. Each nutlet or bur contains 2 to 5 seeds, and has 2 vicious spines on its back. Most of the burs are turned so they lie under the plant. The seeds may remain viable for many years until there is sufficient moisture for germination.

DISTRIBUTION — Puncturevine is native in Europe. It is abundant, one of the most obnoxious weeds in southern Arizona, and is found throughout the state. It is especially troublesome in lawns, citrus orchards, sorghum, alfalfa, cotton, small grain, pastures, gardens, roadsides, yards, and walks; up to 7,000 feet, but mostly at lower elevations; flowering March to October, but principally in July and August.

It appears in remarkable numbers soon after the start of the summer rains on any type of barren soil along the city streets and yards of southern Arizona. In wet summers it is often covered by dodder (*Cuscuta* sp.). Each plant produces innumerable burs, and it is almost impossible to prevent their falling on the ground once they are mature. Home owners and their dogs probably dislike puncturevine more than any other weed because the stout spines can easily penetrate shoes, bicycle tires, and dogs' feet.

Fig. 95. Puncture vine *(Tribulus terrestris).* Prostrate plant with flowers and fruits, or burs, the compound leaves with 4 to 7 pairs of leaflets. *a.* Flower. *b.* Fruit or seedpod, a cluster of five bony burs or nutlets. *c.* Single bur or nutlet containing 2 to 5 seeds.

# NEW MEXICO COPPERLEAF

SPURGE FAMILY — Euphorbiaceae

## NEW MEXICO COPPERLEAF — *Acalypha neomexicana* Muell. Arg.

DESCRIPTION — A weak annual 3 inches to 2 feet high, reproducing only by seed. The bright green soft, lanceshaped leaves are alternate, shallowly toothed on the margins, and 1 to 5½ inches long, including the slender stalks which are ½ to 2½ inches long. The entire plant often turns a copper color in the fall.

The two types of flowers lack petals, and are arranged in terminal or axillary spikes. The small male flowers with 6 to 8 stamens, and the female flowers with the 3 styles, divided into numerous long threadlike parts, are found on the same plant, or each may occur on separate plants. Beneath the female flowers are leaflike bracts conspicuously veined, with the middle tooth elongated. The 3 lobed seedpods are about 1/12 inch long, and somewhat broader. The 3 plump granular, reddish or dark brown seeds are sometimes brown spotted, eggshaped, and about 1/16 inch long.

DISTRIBUTION — New Mexico copperleaf is a native weed growing around drainage ditches, buildings, flower beds, edges of lawns, waste places, roadsides, and overgrazed areas on ranges, also found on shaded slopes and canyons in the mountains, Greenlee to Yavapai county and southward; 2,400 to 7,500 feet elevation; flowering August to October or November.

Fig. 96. New Mexico copperleaf
(*Acalypha neomexicana*). Flowering plant with
the very inconspicuous male flowers in the same
spike as the female flowers, the leaves simple.
*a*. Female flower or young fruit with three
branching styles, subtended by the conspicuous
bract. *b*. Seed.

# SPURGES

SPURGE FAMILY — Euphorbiaceae

POISONOUS PROPERTIES OF SPURGES — All spurges contain a white, sticky milky juice which may cause skin inflammation in some humans and livestock. Livestock rarely eat these green unpalatable plants, but they may eat them in hay. Cattle have been poisoned from feeding on such hay over a period of time. Losses in Arizona from this plant are probably common, but inflammation of the mouth area is more common.

## WHITEMARGIN SPURGE, rattlesnake weed — *Euphorbia albomarginata* Torr. & Gray

DESCRIPTION — A prostrate hairless perennial, forming mats to 3½ feet in diameter, reproducing by seeds and by frequent roots arising at the stem joints. The opposite, hairless leaves, ⅛ to ⅜ inch long, are smooth edged, often edged with white, and red blotched in the center. At the base of the leaf stalk, there is a thin whitish conspicuous scale formed by the 2 united stipules. The 3-lobed seedpods are hairless and each contains 3 seeds. The whitish 4-angled seeds are about 1/16 inch long, with almost smooth faces.

DISTRIBUTION — A common native weed throughout most of Arizona, and especially abundant in the southern part of the state. Found on dry barren disturbed soil along sidewalks, paths, waste places, roadsides, overgrazed or eroded areas around corrals, reservoirs, bedding places, washes, and mesas; 100 to 7,000 feet elevation; flowering February to November.

## SAWTOOTH SPURGE — *Euphorbia serrula* Engelm.

DESCRIPTION — Sawtooth spurge also forms mats, but is annual. The stems have long spreading hairs, and the pale green leaves, either hairless or with a few long hairs, are sharply saw toothed along the edges. The 2 whitish stipules at the base of the leaf stalk are inconspicuous. A native weed with the same general distribution and habitat as whitemargin spurge, but mostly between 2,400 to 8,000 feet elevation; flowering from May to November.

## LITTLELEAF SPURGE — *Euphorbia micromera* Boiss.

DESCRIPTION — A short hairy or hairless, matforming annual, but the leaves are very small, 1/16 to ¼ (mostly ⅛ or less) inch long, and the edges are smooth. The 2 whitish stipules at the base of the leaf stalk are triangular. A native weed found in the same type of places and the same general distribution of whitemargin spurge; flowering August to November.

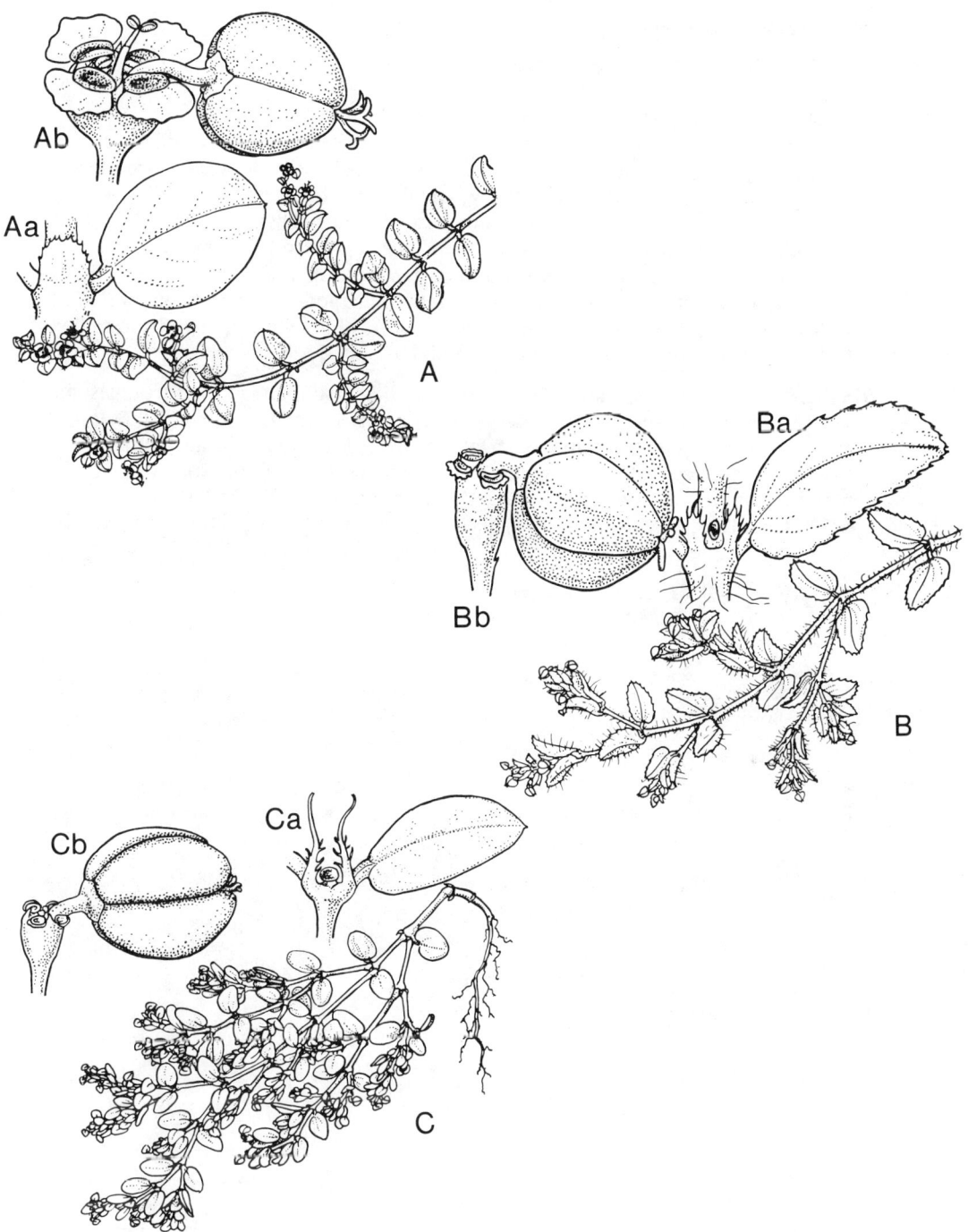

Fig. 97. *A.* Whitemargin spurge *(Euphorbia albomarginata).* Branch with flowers and fruits, the leaves opposite. *Aa.* Leaf with one white membranous scale at base of stalk. *Ab.* Seedpod, also showing four petaloid appendages and four glands of involucre. *B.* Sawtooth spurge *(E. serrula).* Flowering branch. *Ba.* Sawtoothed leaf with two inconspicuous stipules at base. *Bb.* Seedpod, also showing very small glands and petaloid appendages. *C.* Littleleaf spurge *(E. micromera).* Flowering plant. *Ca.* Leaf, with two triangular stipules. *Cb.* Seedpod, also showing lack of petaloid appendages on the involucre.

# SPURGES — continued

SPURGE FAMILY — Euphorbiaceae

## HYSSOP SPURGE — *Euphorbia hyssopifolia* L.

DESCRIPTION — A slender erect hairless annual, often with many stiff spreading branches, 4 inches to 4 feet high in rich soil, reproducing only by seeds. The narrow oblong leaves are opposite, and vary greatly in size. A few leaves, scattered throughout the plant, are at least twice the size of others. They are ⅛ to 1½ inches long, and ⅛ to ⅔ inch broad, with short but sharp pointed teeth on the edges.

The 2 whitish inconspicuous stipules at the base of the leaf stalks are finely shredded along the edges. The greenish 3-lobed seedpods are hairless, about 1/12 inch long, and contain 3 seeds. The dark brown oblong seeds, slightly less than 1/16 inch long, are 4-angled: 3 sharp and 1 rounded.

DISTRIBUTION — This native erect spurge is the most abundant and troublesome spurge in southern and central Arizona. It is a pest from late spring to late fall in all kinds of cultivated crops in the irrigated valleys. It is also a nuisance in gardens, lawns, waste places, often occupying large areas on deteriorating ranges in scrub oak and juniper associations. Frequent along sandy washes and slopes; 100 to 6,000 feet elevation; flowering May to November, mostly August and September.

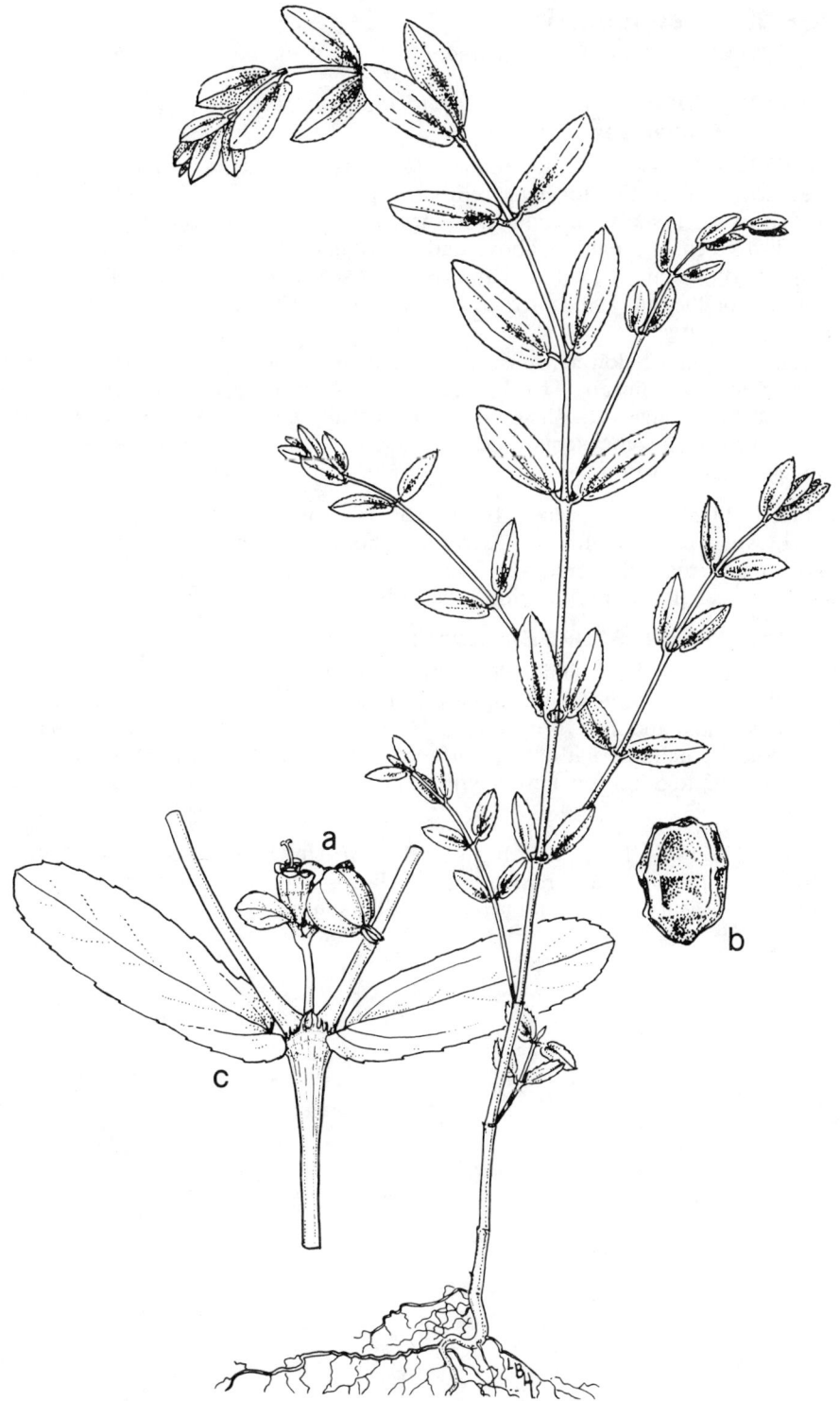

Fig. 98.  Hyssop spurge *(Euphorbia hyssopifolia)*. Leafy plant.  *a.* Seedpod, hairless.  *b.* Seed.
*c.* Leaf, showing two finely shredded stipules at base of stalk.

205

# SPURGES — continued

### GROUNDFIG SPURGE — *Euphorbia prostrata* Aiton
### (*E. chamaesyce* L.)

DESCRIPTION — A prostrate annual often forming mats, or sometimes partially erect where competing for light, reproducing only by seeds. The pinkish to bright red stems have short, spreading, or curved hairs. The opposite leaves, green or purplish green, are hairless above and thinly hairy beneath, ⅛ to ⅓ inch long, with smooth or finely toothed margins. The 2 scalelike stipules, usually present at the base of the leaf stalks, are lacking or very small and inconspicuous in most species of spurge.

The tiny pinkish flowers, consisting of stamens and pistils only, are grouped into small flowerlike clusters. The 3-lobed seedpods are 1/16 inch or less long, hairy only on the 3 angles, the hairs short and usually stiffly spreading or lying close to the surface on some plants. The oblong seeds, about 1/25 inch long, are sharply 4-angled, and have 6 to 8 sharp cross wrinkles on each face.

DISTRIBUTION — Naturalized from Tropical America, groundfig spurge is one of the worst pests in summer lawns throughout southern Arizona. Also common along paths, roadsides, cracks in sidewalks, streets, gardens, and flower beds; 100 to 5,500 feet elevation; flowering June to November.

### PROSTRATE SPURGE — *Euphorbia supina* Raf.

DESCRIPTION — Very similar to groundfig spurge, but the leaves usually have a red blotch in the center. The stipules at the base of the leaf stalks are branched, threadlike, and about 1/16 inch long. The seedpods and seeds are about the same size, but the hairs lie flat on the surface, and are almost evenly distributed. The seeds have 3 to 5 indistinct cross wrinkles on each face, rather than 6 to 8 distinct ones.

DISTRIBUTION — Introduced from the eastern United States, prostrate spurge is also an aggressive and persistent pest in lawns and flower beds, with the same habitat and general distribution of groundfig spurge.

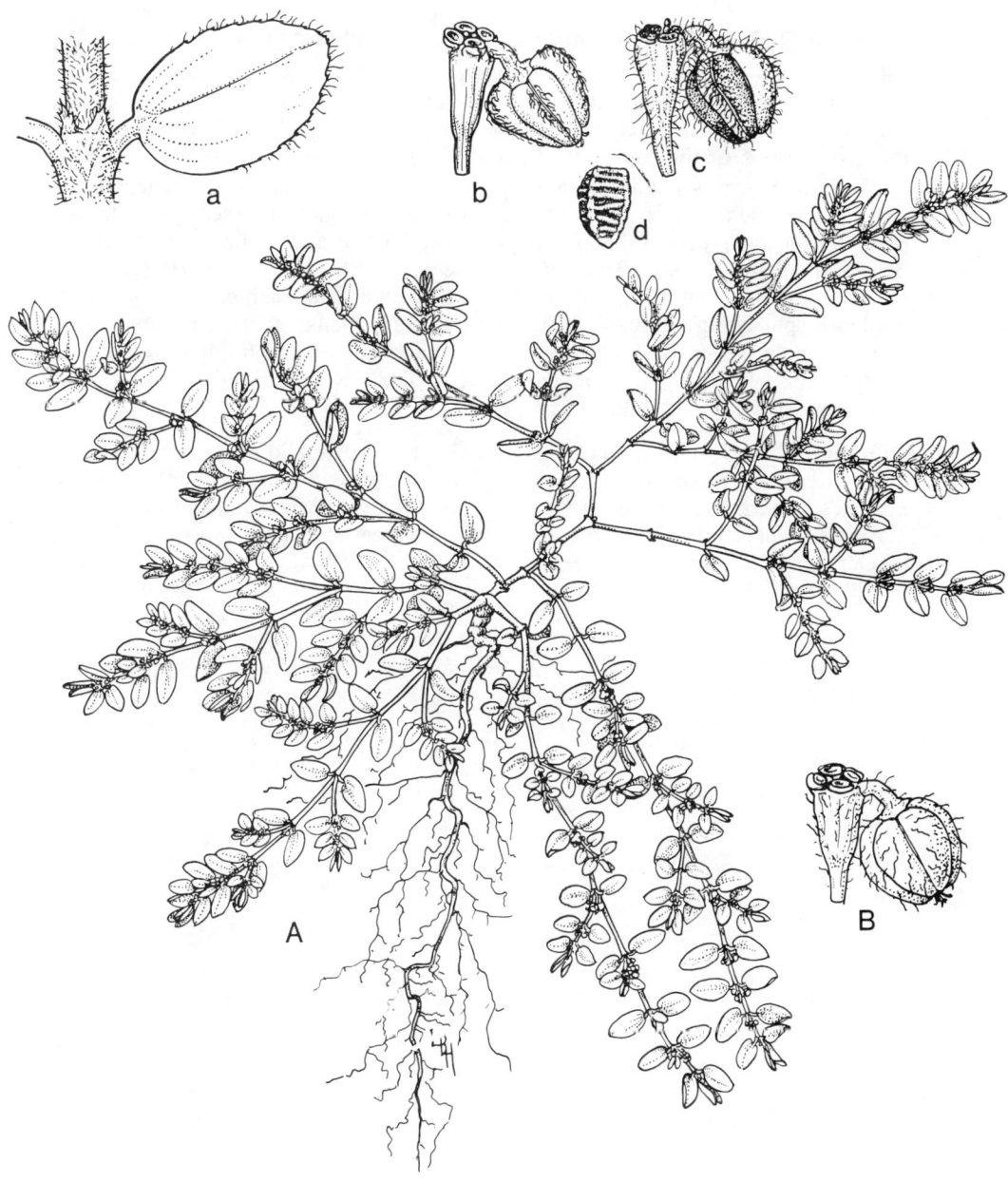

Fig. 99.  Groundfig spurge *(Euphorbia prostrata)*. *A.* Prostrate plant.  *a.* Leaf, showing pair of very small stipules at stalk base, these often lacking completely.  *b.* Seedpod with hairs lying close to the surface.  *c.* Seedpod with hairs stiffly spreading. In both forms, hairs are on the ridges only.  *d.* Seed with sharp cross wrinkles.  *B. Euphorbia supina* capsule.

# SPURRED ANODA

MALLOW FAMILY — Malvaceae

## SPURRED ANODA — *Anoda cristata* (L.) Schlecht.

DESCRIPTION — An erect branched annual ¼ to 3½ feet high, reproducing by seeds. The leaves are alternate, with stalks 1½ to 3 inches long. The blades are triangular in outline, and may be somewhat arrowheadshaped, shallowly lobed with toothed margins, or the basal ones sometimes divided into several fingerlike lobes.

The large flower is solitary on a slender stalk arising at the base of the leaf stalk, and has 5 purple or bluish violet petals ¾ to 1 inch long. The 5 green outer flower parts (the calyx) persist, their long lobes widely spreading under and greatly beyond the flattened disk of the fruit. On the disk are 9 to 20 fruit parts (carpels) which separate at maturity, each forming a seed-like pod. Each carpel has a dark hardened spur along the back, sharp pointed and spinelike at the base, extending beyond the tip into a stiff bristle, 1/12 inch or more long, and thinly covered with conspicuous yellowish hairs. The dark gray seeds are wedgeshaped, about ⅛ inch long, and prickle tipped at the narrow end.

DISTRIBUTION — Spurred anoda is a native weed growing in moist soil, in cotton fields and other irrigated crops, gardens, ditches, and roadsides, also along streams and meadows, mostly in eastern central and southern Arizona, from Apache to Yavapai county and southward; 2,400 to 6,500 feet elevation; flowering August to October or November.

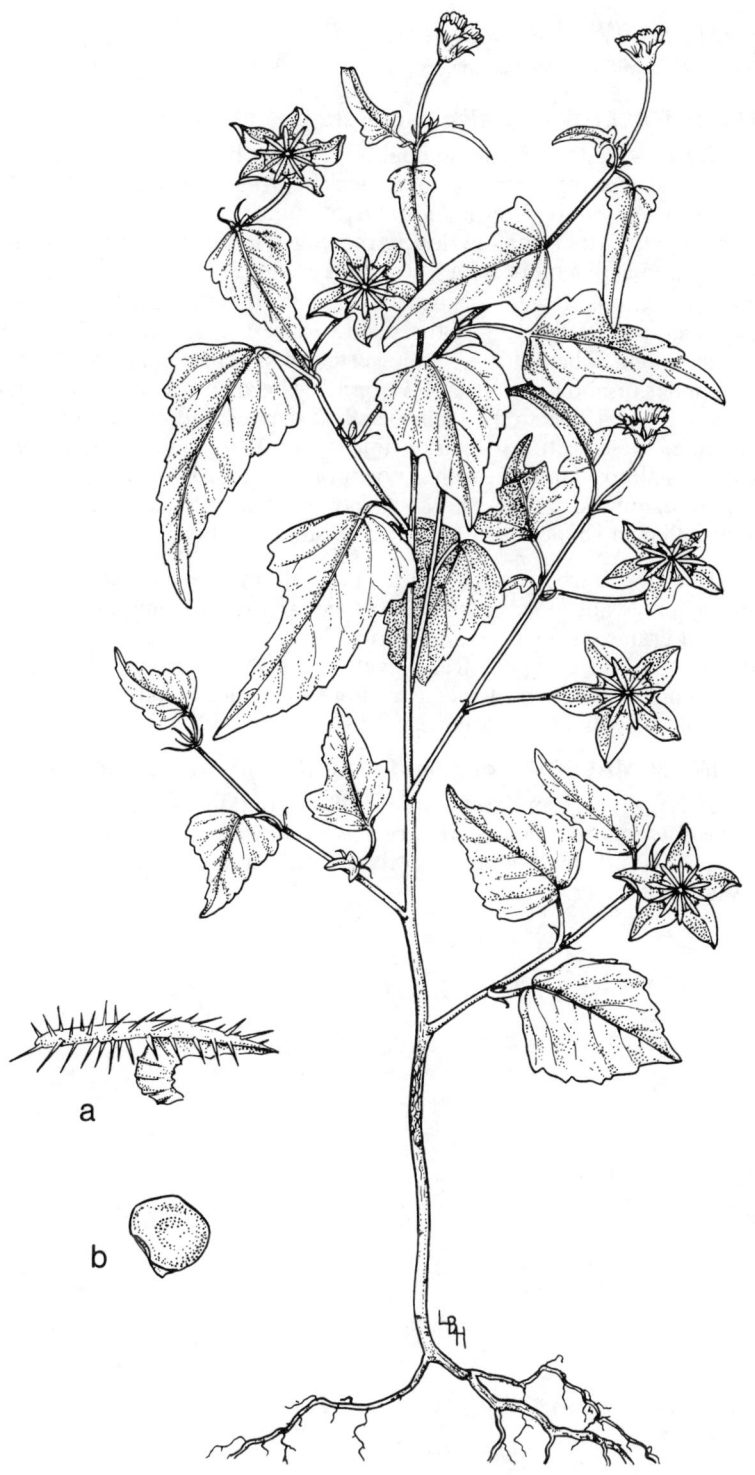

Fig. 100.  Spurred anoda *(Anoda cristata)*. Leafy plant with flowers and fruits, showing enlarged calyx extending beyond the fruit disk; also, top view of the spurs.  *a.* Side view of a carpel with spur on the back.  *b.* Seed.

# LITTLE MALLOW, cheeseweed

MALLOW FAMILY — Malvaceae

### LITTLE MALLOW — *Malva parviflora* L.

DESCRIPTION — A stout, bushy annual or biennial, branched and spreading from the base, 1 to 3 feet high, from a short thick taproot, reproducing only by seeds. The large soft leaves are alternate, almost circular, often with a red spot at the base, and usually 5 to 7 shallow lobes, the edges toothed. The leaves are 2 to 5 inches broad, on stalks more than 4 to 10 inches long.

The inconspicuous flowers are nearly stalkless, in small clusters at the base of the leaf stalks, and at the top of the plant. There are 5 bluish or pinkish petals about ¼ inch long. The outer green flower parts or sepals persist with the 5 lobes spreading under or about the disk of 11 or 12. This disk resembles a round cheese divided into 11 or 12 sections, from which it derives one of its common names. When mature, these sections separate into 11 or 12 seedlike sections or carpels. Each carpel is almost circular, with a notch on 1 edge, one-seeded, flattened, with radiating ridges on the 2 sides and a sharply roughened back. The reddish brown seeds are similar in shape, but the surface is smooth, about 1/12 inch long.

DISTRIBUTION — Introduced from Europe, little mallow is common throughout the state in somewhat moist loam soil. Especially troublesome in irrigated lands, alfalfa, small grains, citrus orchards, yards, ditchbanks, flower beds, waste places, and city streets; 100 to 8,500 feet elevation; flowering the year around, mostly November until June at the lower elevations. The plant and seeds are reported to cause pink egg whites when eaten by hens.

### COMMON MALLOW, roundleaf mallow — *Malva neglecta* Wallr.

DESCRIPTION — Very similar to little mallow, but the stems are spreading or nearly prostrate, and the petals are larger, ⅓ to ⅔ inch long. Found in the same type of places as little mallow, but mostly in northern and central Arizona; flowering May to September.

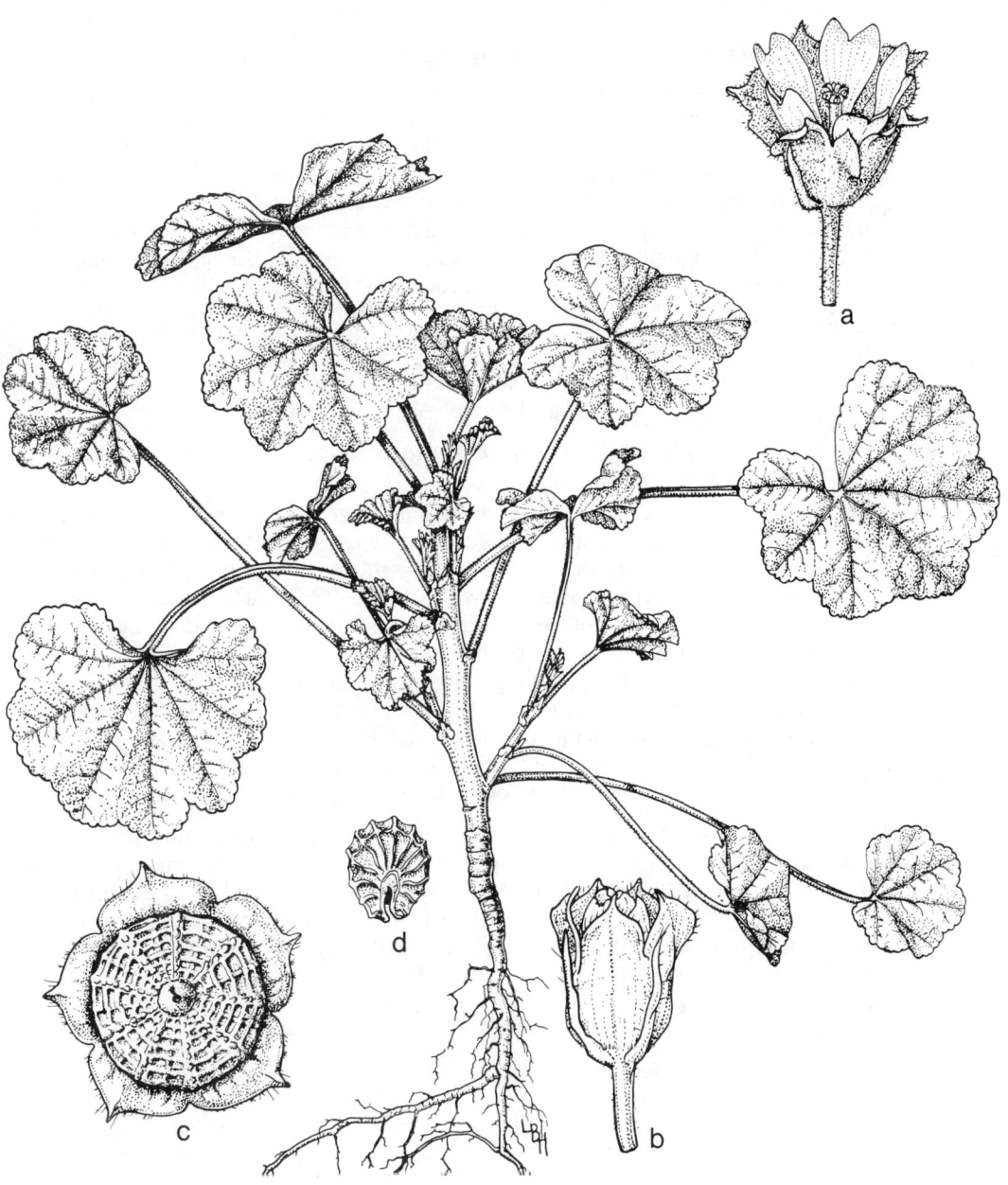

Fig. 101. Little mallow *(Malva parviflora)*. Leafy plant. *a.* Normal flower. *b.* Flower unopened, when temperatures are low. *c.* Fruit surrounded by enlarged sepals. *d.* Carpel with seed enclosed.

# TUBEROUS SIDA

MALLOW FAMILY — Malvaceae

### TUBEROUS SIDA — *Sida physocalyx* Gray

DESCRIPTION — A low perennial which reproduces only by seeds. The branches are nearly erect or spreading at the base and reclining on the ground, ½ to 2 or 3 feet long. The tap root is spindleshaped and tuberlike, about 3 or 4 inches long. The alternate leaves are lanceshaped or broadly oblong, deep green and hairy with toothed margins, stalked. The leaves and stems are lightly covered with branched hairs which are typical of the mallow family. The cream yellow flowers arise from the leaf axils. The outer green flower parts or calyx persist, become much enlarged, papery, 5-angled, winglike, and enclose the fruit.

DISTRIBUTION — Tuberous sida is a common native plant in rich soil in canyons; frequently a nuisance in yards, flowerbeds, and cultivated fields in southern and central Arizona, from Cochise, Santa Cruz, and Pima counties, also in Yavapai County, to 5,000 feet elevation; flowering March to October.

### ALKALI SIDA, alkali mallow, creeping mallow — *Sida hederacea* (Doug.) Torr.

DESCRIPTION — A restricted noxious weed in Arizona, alkali sida is a low whitish or yellowish perennial, densely covered by short yellow, forked hairs, reproducing by seeds, but mostly by long slender woody rhizomes and branched roots. The stems are partially erect to prostrate, ½ to 1⅓ feet long. The leaves are kidney-shaped or rounded with toothed margins, ¾ to 2 inches broad on stalks ½ to 1 inch long. The leaves and stems are covered with scalelike and forked hairs.

The flowers are cream colored when fresh, often fading pink. They occur singly in the leaf axils, and are ⅜ to ¾ inch long. The 6 to 10 dark brown, triangular fruit parts (carpels) are arranged in a disk, each containing a single seed. The seeds are short and kidneyshaped, with one lobe rounded and the other pointed. The seeds seldom mature, because of insect attacks.

DISTRIBUTION — Alkali sida is a native plant. It is often troublesome in cultivated lands and dry ditch banks, particularly in heavy alkaline, bottomland soil, but is not limited to that type of soil. It can be a serious pest in grain and cotton fields, and in orchards. Common from Apache to Coconino county and southward to Graham, Pinal, and Yuma counties; 100 to 5,000 feet elevation; flowering April to October or November.

212

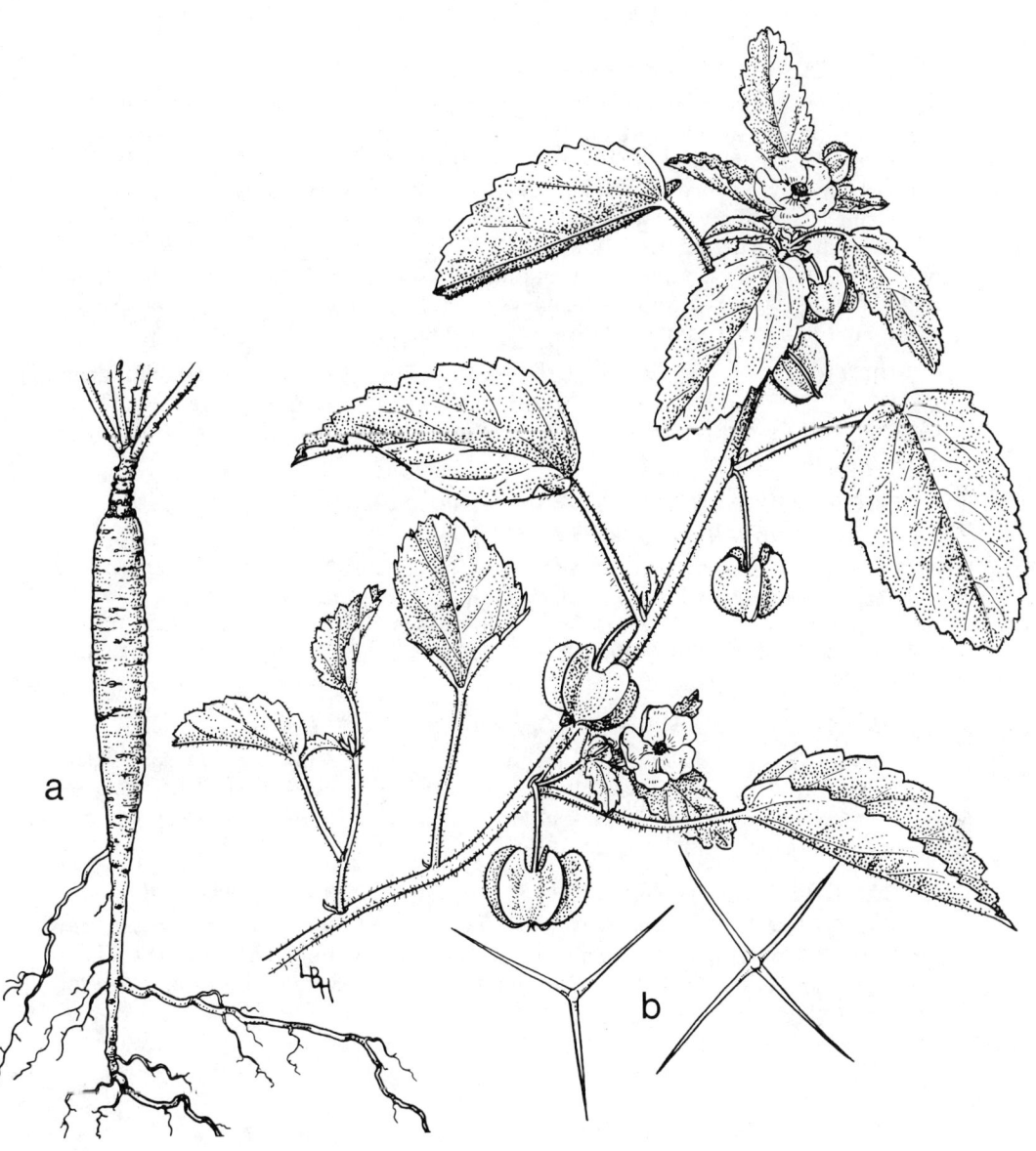

Fig. 102. Tuberous sida *(Sida physocalyx)*. Branch with flowers and fruits, showing the enlarged winglike calyx enclosing fruit. *a*. Taproot. *b*. Forked hairs.

# GLOBEMALLOW

MALLOW FAMILY — Malvaceae

### GLOBEMALLOW — *Sphaeralcea*

DESCRIPTION — Some of the 16 species of globemallow native to Arizona are difficult to distinguish. They are bushy, somewhat woody perennials (two annuals), 1 to 4 feet high, reproducing by seed. The plants are covered by minute starshaped hairs, which may be very irritating to the eyes. The leaves are alternate and palmately veined. The showy flowers have 5 petals mostly shades of pomegranite red, and many stamens united below into a tube. The globeshaped seedpod, surrounded by the persistent calyx, separates at maturity into 7 to 22 kidneyshaped sections. Each section produces 1 to 3 tiny kidneyshaped seeds about 1/16 inch long, dark brown, and more or less hairy with spreading minute hairs.

DISTRIBUTION — Several Arizona globemallows are largely confined to roadsides, borders of cultivated lands, fields, sidewalks, vacant lots, and drainage areas. The perennial species flower about March to May and reflower from August to frost, in response to the summer rains.

### EMORY GLOBEMALLOW — *Sphaeralcea emoryi* Torr., and var. *variabilis* (Cock.) Kearn.

DESCRIPTION — Forms many-stemmed clumps. Leaves varying exceedingly from not lobed to distinctly 3-lobed, to separated into 3 divisions. Petals are usually red, but often pink or lavender, ⅜ to ¾ inch long. Abundant in southern Arizona in Mohave, Graham, Cochise, Pima, Pinal, Maricopa, and Yuma counties; usually below 2,500 feet elevation.

### ORCUTT GOBEMALLOW — *Sphaeralcea orcuttii* Rose

DESCRIPTION — Annual or biennial with a large taproot. The leaves are somewhat triangular with 2 shallow basal lobes. The petals are flame-scarlet to orange, ⅜ to ½ inch long. Found only in southern Yuma County where it is abundant, replacing Emory globemallow; up to 500 feet elevation.

### CALICHE GLOBEMALLOW — *Sphaeralcea laxa* Woot. & Standl.

DESCRIPTION — Leaves shallowly to deeply 3-lobed. Flowers red, the stamen tips (anthers) dark purple. Found on lime soil, especially abundant on caliche mesas around Tucson. Navajo to Coconino, Cochise, and Pima counties; 2,000 to 6,000 feet elevation.

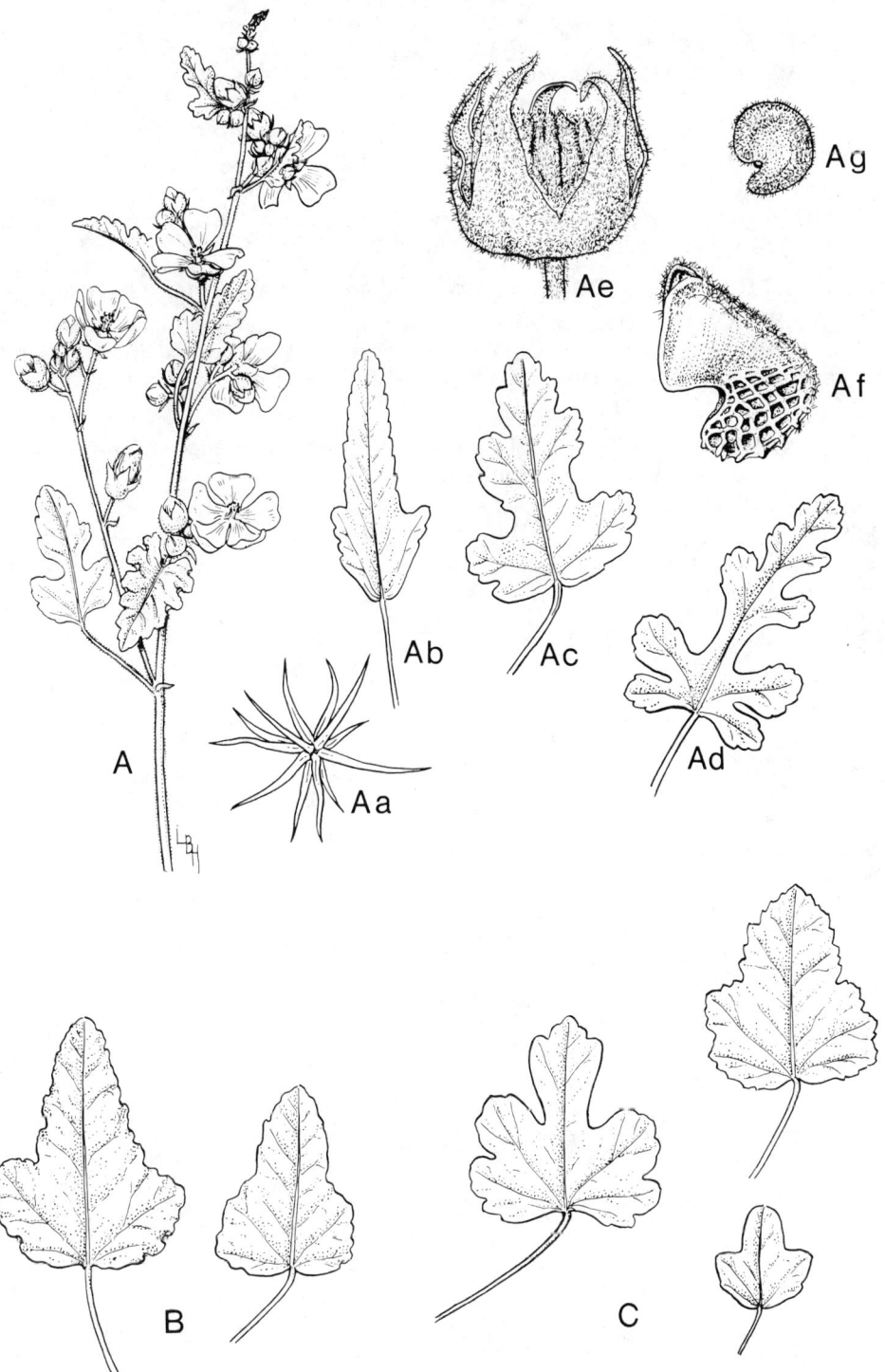

Fig. 103a. Emory globemallow *(Sphaeralcea emoryi* var. *variabilis). A.* Leafy stem with flowers and buds. *Aa.* Starshaped hair, x 20. *Ab., Ac., Ad.* Typical leaves showing variation in shapes. *Ae.* Fruiting calyx. *Af.* One section of seedpod. *Ag.* Seed. *B.* Orcutt globemallow (*S. orcuttii).* Leaf types. *C.* Caliche globemallow *(S. laxa).* Leaf types.

215

# GLOBEMALLOW

MALLOW FAMILY — Malvaceae

## COULTER GLOBEMALLOW — *Sphaeralcea coulteri* (Wats.) Gray

DESCRIPTION — Annual with a slender taproot. Leaves round-triangular, scarcely lobed to deeply 3-to 5-lobed. Petals orange or yellow, ⅜ to ⅝ inch long. Found often in sandy soil; southwestern Arizona, Maricopa, Pinal, Pima, and Yuma counties; below 2,500 feet elevation.

## LITTLELEAF GLOBEMALLOW — *Sphaeralcea parvifolia* A. Nels.

DESCRIPTION — Leaves broadly heartshaped, often shallowly 3-lobed, about ½ to 1½ inches long and broad. Petals red. Abundant in northern Arizona, sometimes in dense patches and grazed by livestock. Apache to Mohave, Yavapai, and Gila counties; 3,500 to 7,200 feet elevation.

## NARROWLEAF GLOBEMALLOW — *Sphaeralcea angustifolia* (Cav.) G. Don var. *cuspidata* Gray

DESCRIPTION — Leaves narrowly lanceshaped, often slightly 2-lobed at base, 1 to 4 inches long, usually about ⅕ as wide. Petals red or pink, about ¼ to ½ inch long. Navajo, Coconino, Cochise, Santa Cruz, and Pima counties; 3,000 to 7,000 feet elevation.

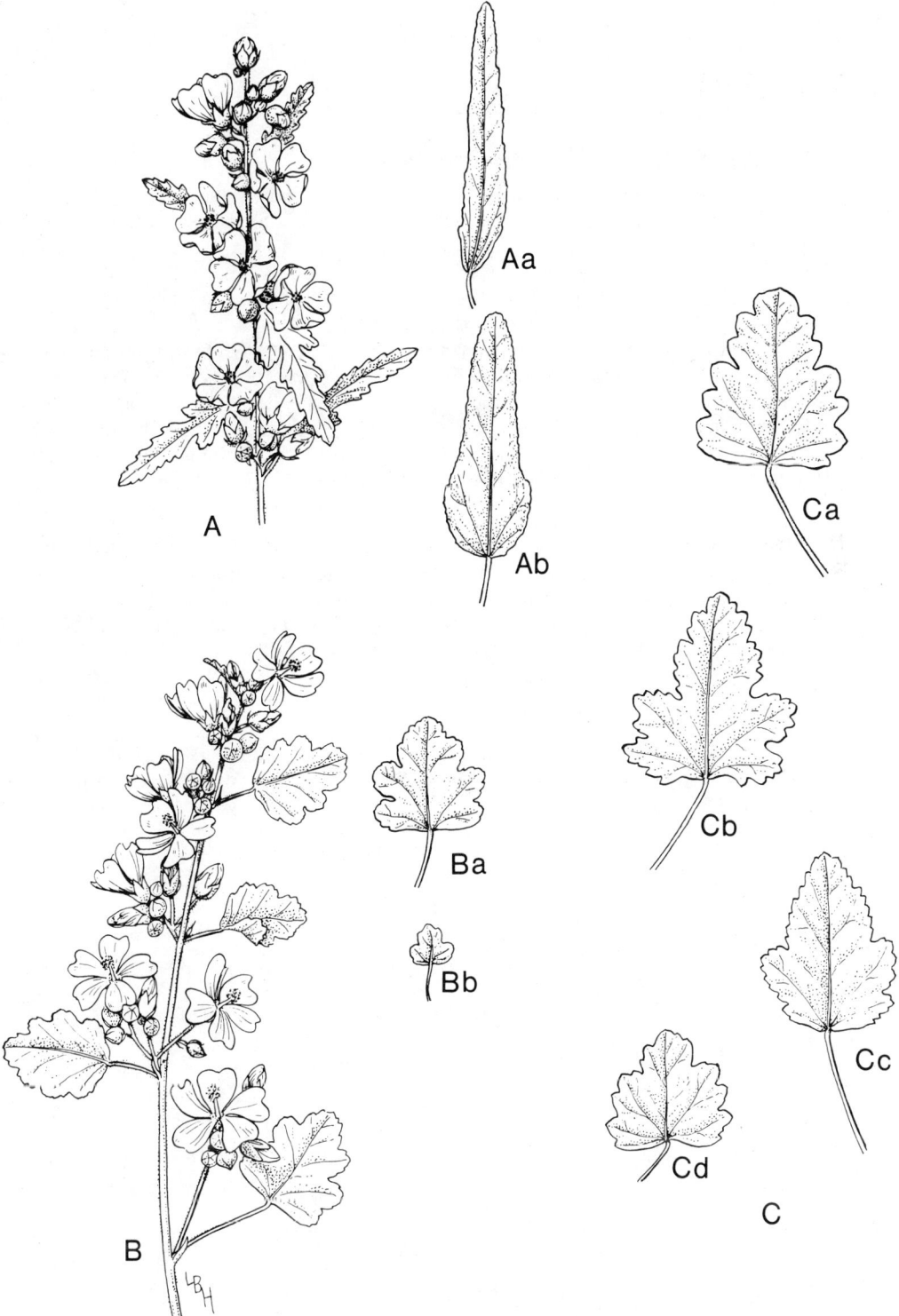

Fig. 103b. *A.* Narrowleaf globemallow *(Sphaeralcea angustifolia* var. *cuspidata).*
Leafy stem with flowers and buds. *Aa., Ab.* Leaf types. *B.* Littleleaf
globemallow *(S. parvifolia).* Flowering stem. *Ba., Bb.* Leaf types. *C.* Coulter
globemallow *(S. coulteri).* *Ca., Cb., Cc., Cd.* Leaf types.

217

## SALTCEDAR, tamarisk

TAMARISK FAMILY — Tamaricaceae

### SALTCEDAR — *Tamarix pentandra* Pall.

DESCRIPTION — A beautiful shrub or small tree, especially in flower, which reproduces by seeds. The long slender grayish green branches are upright or spreading, often forming shrubby thickets many feet in diameter, mostly 10 to 12 (or to 15) feet high. The small grayish green narrow pointed leaves, about 1/16 inch long, are crowded on the stems often overlapping one another. They have the appearance of evergreen leaves, but are actually deciduous.

The myriads of little flowers, from deep pink to nearly white, are about 1/16 inch in diameter and crowded in many slender spikes, ½ to 2 inches long, forming a dense showy mass at the top of the branches. The slender tapering many seeded pods are pinkish red to greenish yellow, ⅛ to nearly ⅕ inch long, splitting into 3 to 5 parts when mature. The tiny seeds are less than 1/25 inch long, and have a tuft of fine silky hairs at the tip. The bark is reddish brown and fairly smooth at first, but becomes ridged and furrowed.

DISTRIBUTION — Saltcedar was introduced from Eurasia, but soon escaped cultivation. Now it is abundant in the bottomlands, banks, and drainage washes of all the rivers and other watercourses throughout the state, including the irrigation ditches and the Grand Canyon. It is very drought-resistant, and grows in all types of soil, including alkali and salty soil. Found principally in deserts and desert grasslands at 100 to 5,000 feet elevation or sometimes to 6,000 feet in northern Arizona, as in Canyon de Chelly (Apache County), where it is common; flowering March to September.

At one time saltcedar was looked upon with favor as a check to erosion, a windbreak, and a source of honey, but now its abundance makes it a nuisance in the big rivers.

218

a

b

Fig. 104. Saltcedar *(Tamarix pentandra)*. Branch showing tiny
leaves and spikes of flowers. *a.* Flower. *b.* Seed with tuft of hair.

219

# SMALLFLOWER GAURA, velvetweed

EVENING PRIMROSE FAMILY — Onagraceae

## SMALLFLOWER GAURA — *Gaura parviflora* Dougl.

DESCRIPTION — A tall upright annual, winter annual, or biennial from a thick taproot, reproducing only by seeds. There is 1 principal stem, 2 to 6 feet high, usually branching above the middle and covered with fine silky stiff hairs. The leaves first form a rosette on the ground. The stem leaves are alternate, nearly stalkless, lanceshaped and soft hairy, 2 to 4 inches long.

The small pink or reddish flowers with 4 petals, 1/16 to 1/12 inch long, are stalkless and crowded on long branching flower spikes at the top of the plant, 2 inches to 18 inches long. The hard woody seedpods are greenish, ¼ to ⅓ inch long, ribbed and usually hairless, or occasionally short hairy. They are spindle-shaped, pointed at both ends, and contain 2 to 4 seeds, which are not shed at maturity.

DISTRIBUTION — Smallflower gaura is a native western weed which is common throughout Arizona in moist soil of cultivated lands, old fields, roadsides, barren areas on overgrazed flats in the higher desert grasslands or sandy river washes; 100 to 6,800 feet elevation; flowering April to October or November.

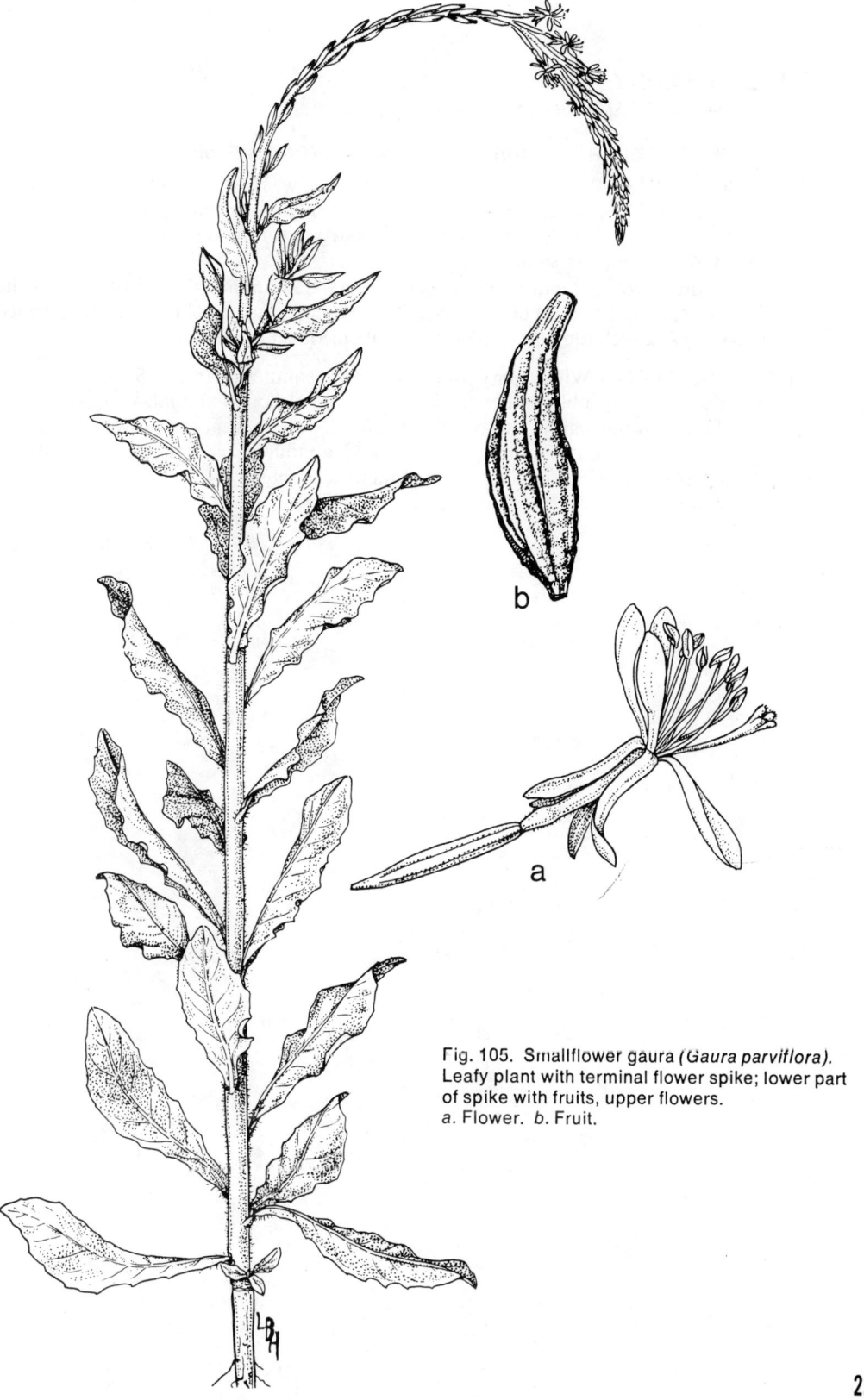

Fig. 105. Smallflower gaura *(Gaura parviflora)*.
Leafy plant with terminal flower spike; lower part
of spike with fruits, upper flowers.
*a.* Flower. *b.* Fruit.

221

# WILD CELERY

PARSLEY FAMILY — Umbelliferae

## WILD CELERY — *Apium leptophyllum* (Pers.) F. Muell.

DESCRIPTION — A slender hairless annual, the weak stems erect or spreading, ⅓ to 2 feet high, which reproduce by seeds only. The delicate alternate leaves are divided and redivided into many very narrow to threadlike segments, and are widely spaced on the stems.

The tiny inconspicuous white flowers are in small umbrellalike clusters opposite the leaves, and at the top of the plant. The flattened oval to oblong seedlike fruits, 1/16 to 1/12 inch long, have 5 or 6 prominent ribs on each face.

DISTRIBUTION — Wild celery is native from southern United States to South America. It is a troublesome weed in lawns, and although annual, very persistent. Constant mowing causes it to branch often, but does not otherwise affect it. Common throughout southern Arizona and abundant in certain areas; 100 to about 5,000 feet elevation; flowering March to September.

Fig. 106. Wild celery *(Apium leptophyllum)*.
Leafy plant with flowers and fruits.
*a*. Flower. *b*. Fruit.

223

# HAIRY BOWLESIA

PARSLEY FAMILY — Umbelliferae

## HAIRY BOWLESIA — *Bowlesia incana* Ruiz & Pavon

DESCRIPTION — A delicate annual with short forked hairs, branching and flowering from the base, reproducing only by seeds. The weak stems are nearly erect or trailing on the ground, ¼ to 2 feet long. The thin opposite leaves are mostly 5-lobed (or 3 to 7), broader than long, ½ to 1 inch broad, on slender stalks ½ to 3 inches long.

One to 4 tiny greenish white flowers occur at the base of the leaf stalks. The rounded seedlike fruits are about 1/16 inch long, with no ribs on the faces.

DISTRIBUTION — Hairy bowlesia is a native weed occurring in shady places in moist or dryish, often barren soil around buildings, yards, waste places, and under bushes or rocks in sandy washes, desert mesas and slopes, from Mohave and Coconino counties southward; 100 to 3,500 feet elevation; flowering January to May or June.

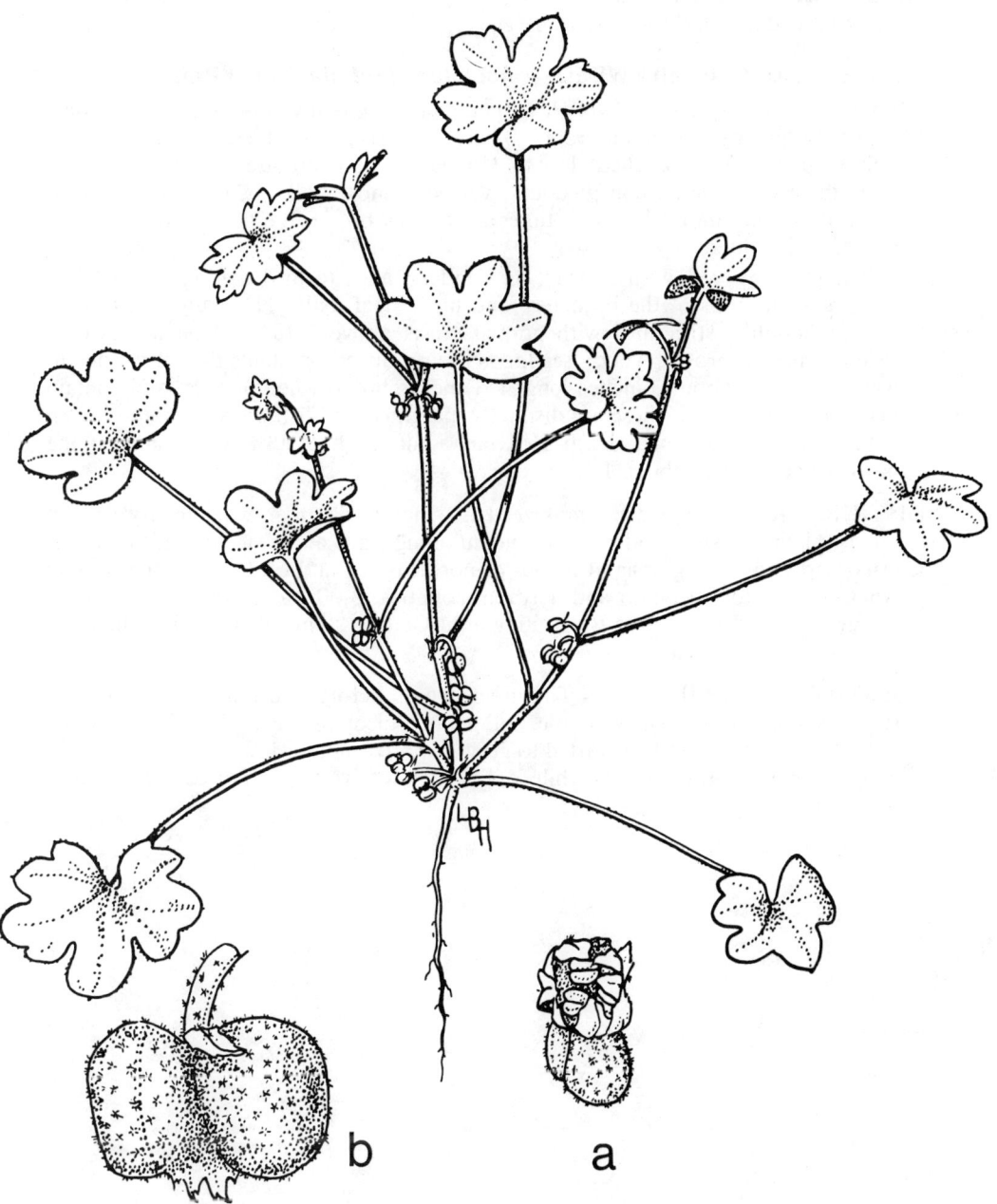

b       a

Fig. 107. Hairy bowlesia *(Bowlesia incana).* Leafy plant with
tiny flowers and fruits at the leaf bases. *a.* Flower. *b.* Fruit.

# BROADLEAF MILKWEED

MILKWEED FAMILY — Asclepiadaceae

## BROADLEAF MILKWEED — *Asclepias latifolia* (Torr.) Raf.

DESCRIPTION — A robust, very leafy perennial with milky juice, which reproduces by seeds and by horizontal roots. There is 1 stout, erect stem, often hollow, unbranched or few branched, 1½ to 3½ feet high. There are 5 or more pairs of large thick oval leaves, conspicuously veined, almost stalkless, often as broad as long, and rarely more than 1½ times as long as broad, the tip broadly rounded and often indented.

The fragrant, yellowish flowers, on slender stalks, form drooping umbrellalike clusters at the tops of the branches, and in the leaf axils. These unusual flowers have 5 hoodlike structures with horned crests above 5 reflexed petals. Two to several smooth, brown, woody seedpods are produced on stalks that curve downward. They are about 3 inches long, 1¼ inches broad when mature, and contain many flattened seeds. These are distinctly margined, reddish brown, about ⅓ inch long, and ¼ inch broad. Each seed has a tuft of fine silky white hairs at the narrow end, and is about 1 inch long.

DISTRIBUTION — Broadleaf milkweed is a native plant growing in dry soil on mesas, plains, washes, and often abundant along trails and roadsides in the northern desert and short grassland ranges of northeastern and central Arizona. Found from Apache to Coconino and Yavapai counties; 3,000 to 7,000 feet elevation; flowering from June to August, fruiting until October. This plant tends to increase in heavily grazed areas.

POISONOUS PROPERTIES — The green plants before and during the flowering stage are poisonous to sheep, cattle, and goats. Poisoning has occurred early in the spring before the grass had started to grow. Approximately 1% of animal's weight of green plant causes death, while only 0.2% of western whorled milkweed is required.

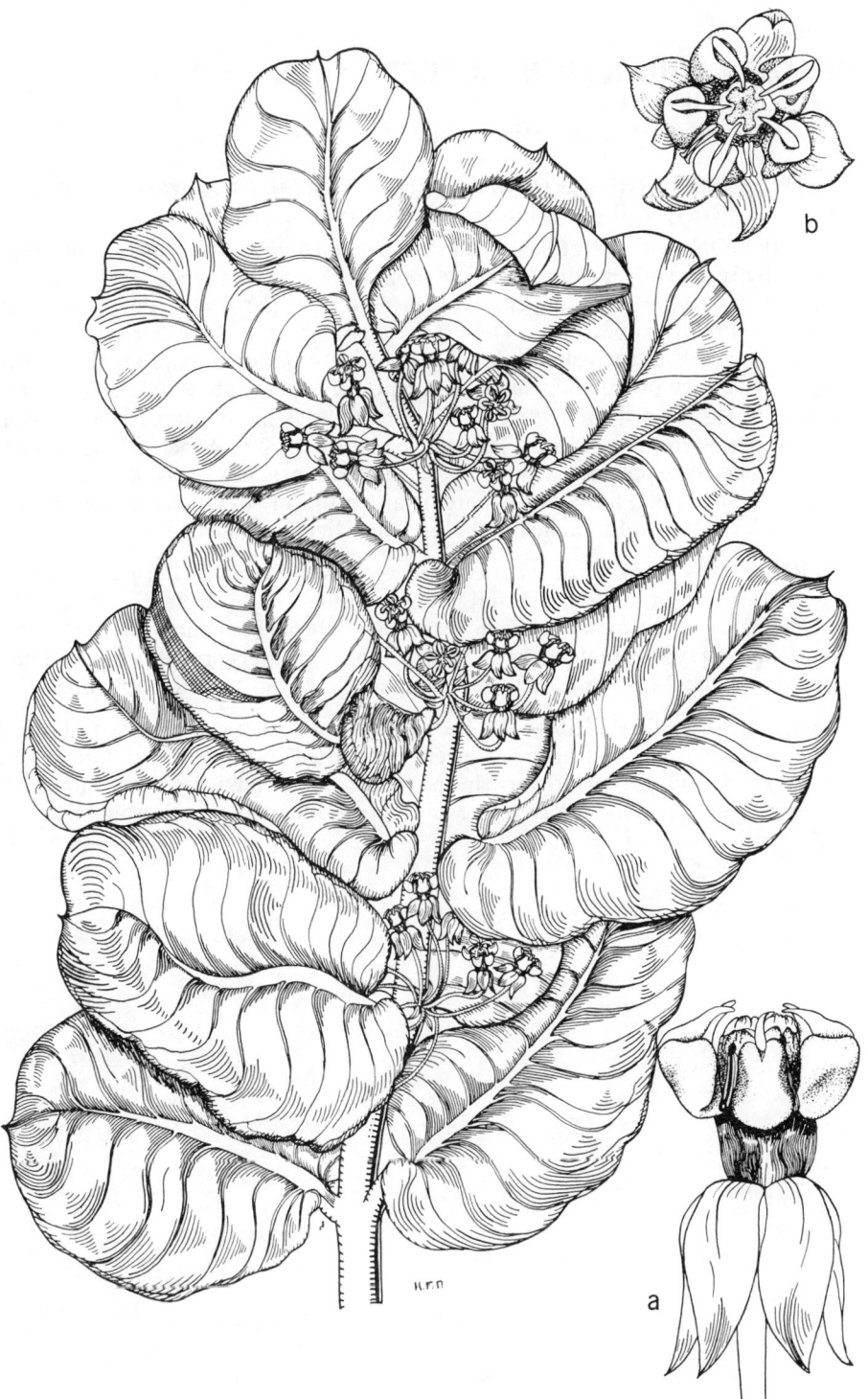

Fig. 108.  Broadleaf milkweed *(Asclepias latifolia).*  Upper part of plant
showing stout stem, large, thick oval leaves, and clusters of flowers.
*a.* Enlarged flower showing reflexed petals, the hoods arising from them
and the horns of the hoods projecting inward.  *b.* Top view of flower.

227

# WESTERN WHORLED MILKWEED, horsetail milkweed, poison milkweed
## MILKWEED FAMILY — Asclepiadaceae

### WESTERN WHORLED MILKWEED — *Asclepias subverticillata* (Gray) Vail (*A. galioides* of authors, not H.B.K.)

DESCRIPTION — An erect bushy, hairless perennial with milky juice, which reproduces by seeds and by wiry, creeping roots. There are many slender unbranched stems 1 to 4 feet high, arising close together from the horizontal roots, giving large plants a shrubby appearance. Three (sometimes 2 or 4) long narrow leaves are found in whorls at each stem joint. These are 3 to 5 inches long, nearly stalkless, not over 1/3 inch broad and usually less, with the edges slightly rolled backwards.

The small, greenish white flowers occur in umbrellalike clusters at the top of the branches, and in the leaf axils. The slender seedpods, 2 to 4 inches long, splindle-shaped and long pointed, with erect stalks, contain many seeds. The flat brown seeds, 1/4 inch long, have corky margins and a tuft of silky hairs at the tip. They are similar to those of broadleaf milkweed, but slightly smaller.

DISTRIBUTION — Western whorled milkweed is native in western United States and Mexico. It grows in dryish soil along roadsides, edges of fields, and pastures, ditchbanks, plains, mesas, and slopes. Found throughout most of the state on practically all types of ranges from desert to spruce fir, it is probably most common in yellow pine; 2,500 to 8,500 feet elevation; flowering May to September.

POISONOUS PROPERTIES — All parts of western whorled milkweed above the ground are poisonous at all times, even when dried. It is poisonous to all classes of livestock, but particularly to sheep. None of the milkweeds are palatable to livestock, and animals will rarely touch them if other forage is available.

Fig. 109. Western whorled milkweed *(Asclepias subverticillata)*. Flowering branch with the leaves in whorls. *a*. Seedpod. *b*. Flower. *c*. Seed with tuft of silky hairs.

# CLIMBING MILKWEED
MILKWEED FAMILY — Asclepiadaceae

## CLIMBING MILKWEED — *Funastrum cynanchoides* (Decne.) Schlecht., *F. heterophyllum* (Engelm.) Standl.

DESCRIPTION — Perennials with milky juice and long twining stems 8 to 40 feet long, which reproduce by seeds and horizontal roots. The leaves are in pairs, thinly covered by short hairs to hairless. The various plants show remarkable variability in the size and shape of the mature leaves. They all have slender stalks, which vary in length from 1/16 inch to over 2 inches. The leaf blades are very narrow — 1/16 inch — to very broad — 1¾ inches. The leaf base may be narrowed to a point and taper into the stalk, or they may be rounded, or 2-angled, or heartshaped, or arrowshaped, or 2-lobed with the lobes pointing outward. They may be narrowed to a long slender point at the tip, or short pointed, or rounded with a point tip. They are ½ to 2½ inches long.

The flowers are in characteristic umbrellalike clusters, with 15 to 25 on a slender stalk arising at the leaf axil. They are fragrant, white, whitish yellow, or purple. The brown seedpods are 1½ to 3½ inches long, ⅓ to ⅔ inch wide. The reddish brown seeds are ⅛ to ¼ inch long, slightly toothed at the rounded end, with a tuft of soft silky hairs at the narrow end.

DISTRIBUTION — Climbing milkweed is a native weed growing in dry sandy soil, and a nuisance around ranches, often climbing on fences, ditches, small trees and shrubs, and spreading into gardens and cultivated fields. Also common in desert washes and arroyos throughout southern and central Arizona; 100 to 5,500 feet elevation; flowering April to October; the seedpods persisting until November.

Fig. 110. Climbing milkweed *(Funastrum cynanchoides* and *F. heterophyllum).* Twining plant with flower clusters and seedpods. *a.* Leaf variation. *b.* Seed. *c.* Flower.

# FIELD BINDWEED

### MORNING GLORY FAMILY — Convolvulaceae

## FIELD BINDWEED — *Convolvulus arvensis* L.

DESCRIPTION — A prohibited noxious weed in Arizona, field bindweed is a prostrate perennial from a thick, branching, deeply penetrating taproot, which reproduces by seeds and by extensive horizontal roots and rhizomes. The slender stems, twining or trailing on the ground, are 1 to 3 (or 10) feet long. The leaves are alternate on stalks ¼ to ¾ inch long. The leaf blades are eggshaped with the base squarish, heartshaped, arrowshaped, or the 2 lobes pointed outward. They are ½ to 1⅞ inches long, ⅔ to 1 inch broad, and usually roundish at the tip.

The funnelshaped flowers are white or pink, ⅔ to 1 inch long, and ¾ to 1 inch across. They occur singly (sometimes 2 or 3) on stalks 1 to 2 inches long in the leaf axils. A pair of narrow pointed bracts ⅛ to ¼ inch long occur on the flower stalks ½ to 1 inch below the flower.

The globeshaped, point-tipped seedpods contain 4 chocolate brown seeds. These are somewhat eggshaped, the surface roughened, 3-angled, or flat on 1 side, rounded on the other, and ⅛ to ⅕ inch long.

DISTRIBUTION — Introduced from Europe, field bindweed grows mostly in dry soil along roadsides, in open fields or edges of cultivated fields, pastures, on fences, yards, and waste places of farms. Also found in alfalfa, small grains, cotton, and sorghum. Common throughout the state; 100 to 8,500 feet elevation; flowering March to November. Considered the most noxious of all weeds in several western and midwestern states, this drought resistant plant with its unusual root system is almost impossible to eradicate from an infested area.

## HEDGE BINDWEED — *Convolvulus sepium* L.

DESCRIPTION — Similar to field bindweed, but the 2 bracts are larger, ½ to 1 inch long and are immediately below the flower. The flowers are larger, 1½ to 2½ inches long, and the leaves also are larger, 2½ to 3½ inches long.

DISTRIBUTION — Not common in Arizona except locally in somewhat moist soil, from Apache to Coconino counties; 6,000 to 7,000 feet elevation; flowering June to August.

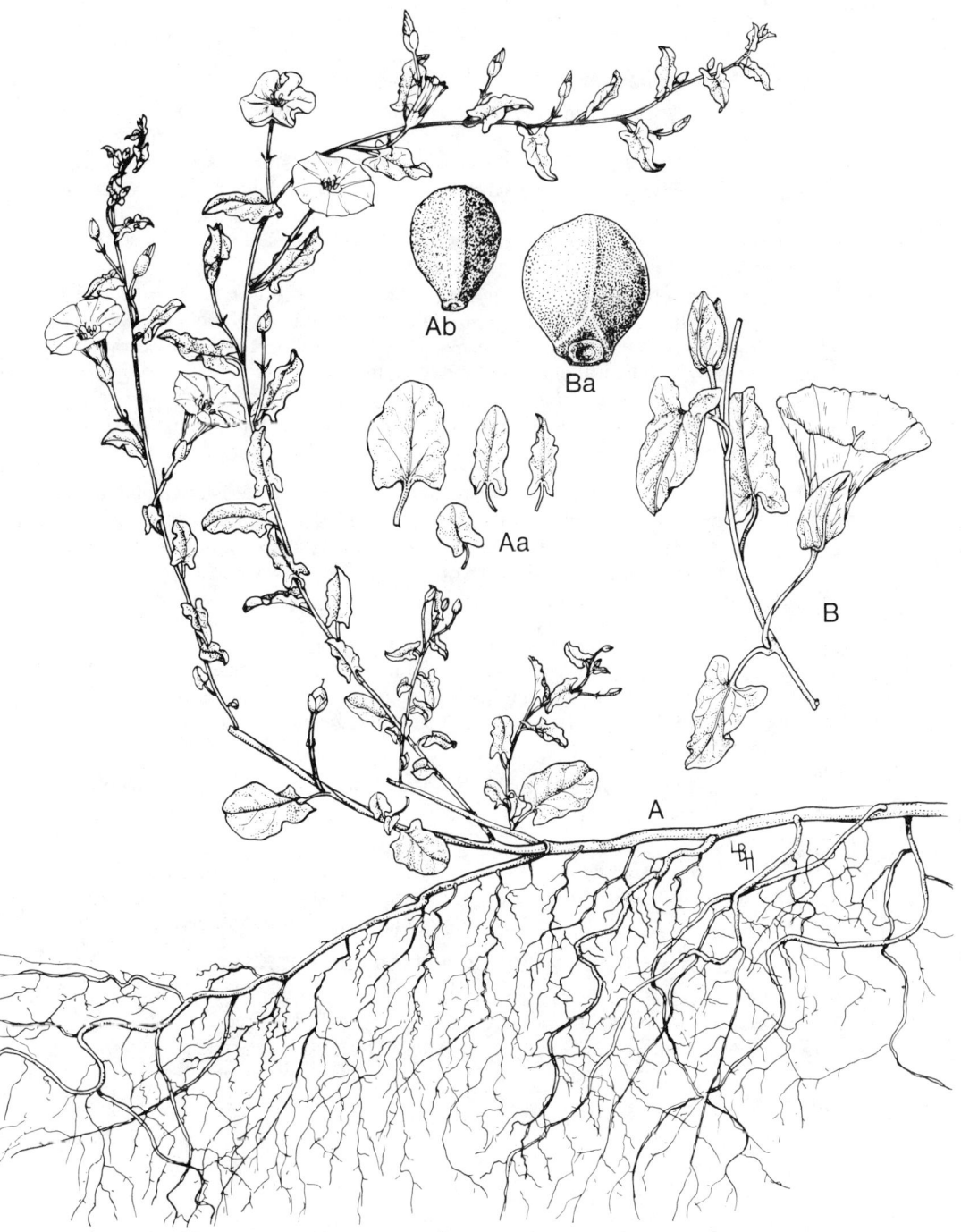

Fig. 111. *A.* Field bindweed *(Convolvulus arvensis)*. Prostrate plant with both flowers and seedpods. *Aa.* Various shapes of leaves. *Ab.* Seed. *B.* Hedge bindweed *(C. sepium)*. Branch with flower. *Ba.* Seed.

# DODDER

MORNING GLORY FAMILY — Convolvulaceae

### DODDER — *Cuscuta* Spp.

DESCRIPTION — A restricted noxious weed in Arizona, dodder is a parasitic, rootless, leafless annual vine with beautiful yellow or orange, stringlike, twining stems, reproducing only by seeds. The leaves are reduced to colorless scales, and the plant has no green matter. Upon germination, the long slender seedling, at first dependent upon food stored in the seed, coils about an available plant to which it becomes attached by numerous suckers. It then depends upon its host for all food; should the seedling fail to reach a host, it soon dies.

The stems branch greatly, forming a network about the host and spread on to the next plant, the growing parts continually producing new suckers. Although branches of the host or the entire host may die, the only part of the vine that dies is that portion directly attached to lifeless material.

The white or cream colored flowers are bellshaped, fleshy, mostly 5-lobed, 1/12 to ¼ inch long, and arranged in clusters along the stems. The globeshaped seedpods are thin and paperish, producing 2 to 4 seeds.

DISTRIBUTION — Dodder is a noxious weed, sometimes parasitizing important crops, particularly alfalfa and clover, with whose seeds it is often mixed. Of the 15 species in Arizona, only 3 are common, ranging from Coconino and Yavapai counties southward; 100 to about 5,000 feet elevation; flowering June to November, but mostly July to September.

### FIELD DODDER — *Cuscuta campestris* Yunck.

DESCRIPTION — The withered flower parts are persistent about the base of the seedpod, and the grayish tan, eggshaped seeds are 1/25 to 1/16 inch long. Growing on many hosts, but particularly on sugar beets, also alfalfa, clover, and other legumes.

### LARGESEED DODDER, bigseed alfalfa dodder — *Cuscuta indecora* Choisy

DESCRIPTION — The withered flower parts completely cover the seedpods; the reddish tan, circular seeds are 1/16 to 1/12 inch at the longest axis. Attacks woody as well as herbaceous plants, often on sunflower, goldenrod, aster, and burrobrush; frequently on legumes such as alfalfa, mesquite, and catclaw.

### UMBRELLA DODDER — *Cuscuta umbellata* H.B.K.

DESCRIPTION — The flowers are in umbrellalike clusters, with the withered flower parts wholly covering the seedpods. The greenish to reddish tan seeds are 1/25 to 1/16 inch long. Especially common on puncturevine, often covering large areas, also on cultivated beets, horse purslane, and various spiderlings and spurges.

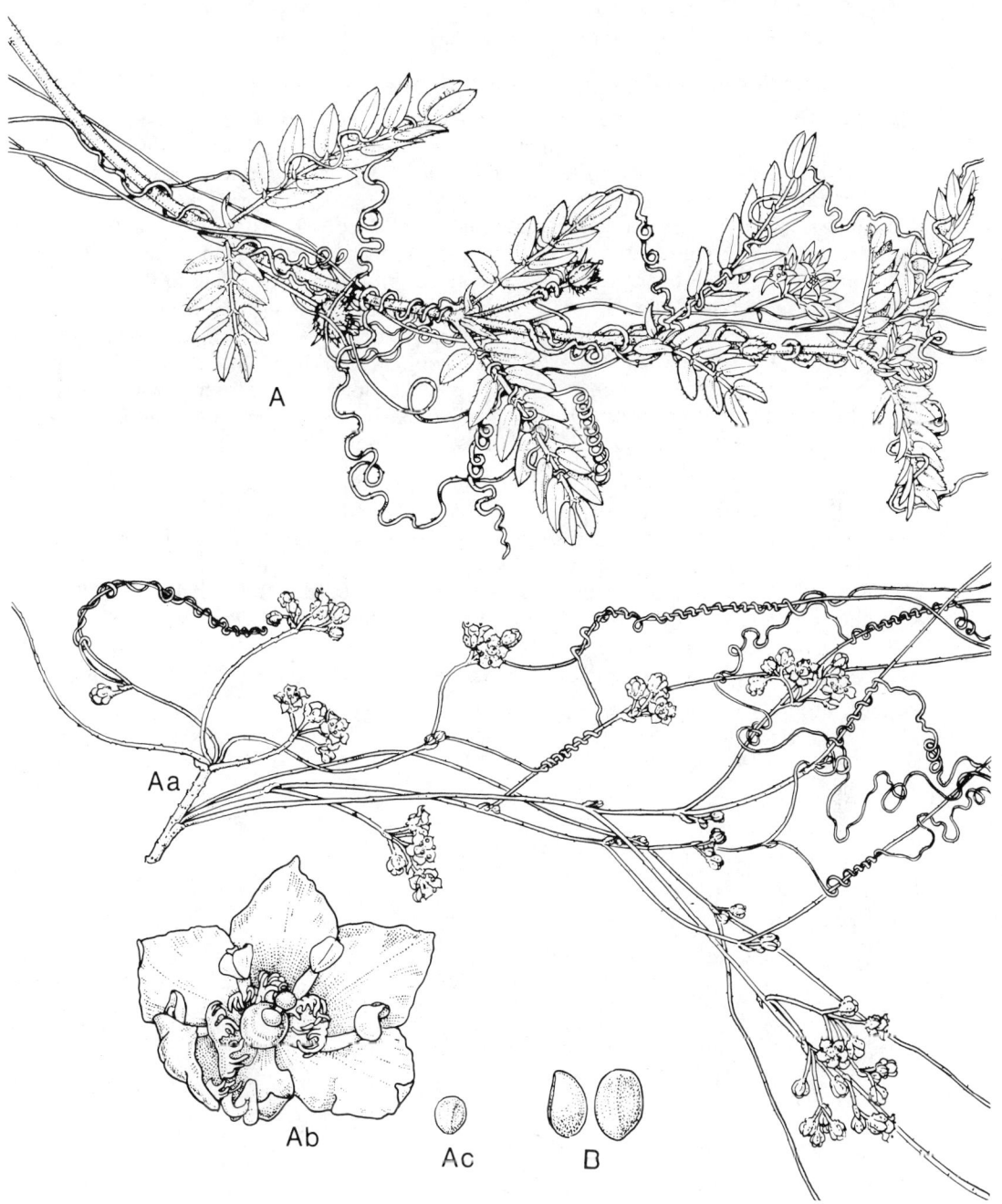

Fig. 112. Umbrella dodder *(Cuscuta umbellata)*. *A.* Vine entwined
on host, orange caltrop. *Aa.* Plant showing flowers and twining
habit. *Ab.* Flower. *Ac.* Seed. *B.* Largeseed dodder *(C. indecora).*

## SCARLET MORNINGGLORY, red morningglory, starglory
MORNING GLORY FAMILY — Convolvulaceae

### SCARLET MORNINGGLORY — *Ipomoea coccinea* L.

DESCRIPTION — All species of morningglory are declared prohibited noxious weeds in Arizona. Scarlet morningglory is a hairless annual with ridged, often reddish stems, twining or trailing on the ground, which reproduces only by seeds. The alternate leaves, on stalks 1 to 4 inches long, are of 2 principal shapes. On some plants they are unlobed, with the base deeply heartshaped and the tip conspicuously long pointed, 1½ to 2½ inches long. On other plants (var. *hederifolia*), some or all of the leaves are deeply cut into 3 to 5 fingerlike lobes.

The scarlet red flowers easily separate scarlet morningglory from all other Arizona species. (Others are pink, purple, blue, or white.) The flowers are narrowly trumpetshaped, 1 to 1¼ inches long, and ½ to ⅔ inch across. Two to several flowers are borne on a stalk 3 to 5 inches long, arising from the leaf axils. The globeshaped seedpods contain 4 to 6 blackish seeds. These are plump, somewhat eggshaped but angular, ⅛ inch or slightly more long.

DISTRIBUTION — Scarlet morningglory is native in Arizona, New Mexico, and in tropical America. It is a pest in cotton fields and other commercial croplands in southeastern Arizona, from Apache to Coconino county and southward. Particularly troublesome in Graham County (around Safford and Thatcher) and in Santa Cruz County. Often found along roadsides, ditches, sandy washes, hillsides, and canyons; 2,000 to 7,000 feet elevation; flowering May to October.

All species of morningglory are serious in cotton fields after the cultivating period is over. They grow unchecked then, and wind about the plant, the flower, and the boll.

Fig. 113. Scarlet morningglory *(Ipomoea coccinea var. hederifolia).*
Flowering branch, with fingerlike lobed leaves. *a.* Seedpod. *b.* Seed.

# WOOLLY MORNINGGLORY, Mexican morningglory
MORNING GLORY FAMILY — Convolvulaceae

## WOOLLY MORNINGGLORY — *Ipomoea hirsutula* Jacq. f.

DESCRIPTION — A prohibited noxious weed in Arizona, woolly morningglory is
an annual with twining or trailing stems, up to 20 feet long, from a taproot system.
All parts are hairy; it reproduces by seeds only. The leaves are of 3 principal
shapes, on stalks 2 to 4 inches long. Some are unlobed, heartshaped, and very
similar to the leaves of tall morningglory. Others vary from barely angulate to
3-lobed, to very deeply 3-lobed, with conspicuously heartshaped bases, 1½ to 4
inches long. A few may be divided into 5 fingerlike lobes.

The flowers are blue, purple or whitish, and as in other morningglory flowers,
open early in the morning, and close soon after the sun shines. They are 1 to 1¾
inches long, and in clusters of 1 to 5 on the long flower stalk. The 5-lobed calyx
is conspicuously hairy at the base; these are ⅓ to ½ inch long, or in some plants
up to 1 inch long.

The globeshaped seedpod is yellowish, and contains 4 seeds. The seeds are
similar to those of scarlet morningglory, but larger, about ¼ inch long, dark
reddish brown to black, minutely hairy, and more flattened.

DISTRIBUTION — A native of tropical America, woolly morningglory is the most
noxious of the Arizona species. Widely distributed throughout the central and
southern part of the state, infrequent northward, it is abundant in cultivated
lands. Especially troublesome in cotton, soybean, sorghum, and corn fields, roadsides, and sometimes in pinyon and yellow pine forests; from 100 to 7,000 feet
elevation; flowering from May to November.

Around St. David (Cochise County) after the onset of the heavy summer
rains, it covers more than 1,000 acres, mostly lying flat on the ground. It is one
of the worst late cotton pests in the Safford (Graham County), Avra, Santa Cruz
(Pima County), and Yuma valleys. Probably the major infestations occur in
Maricopa and Pinal counties.

238

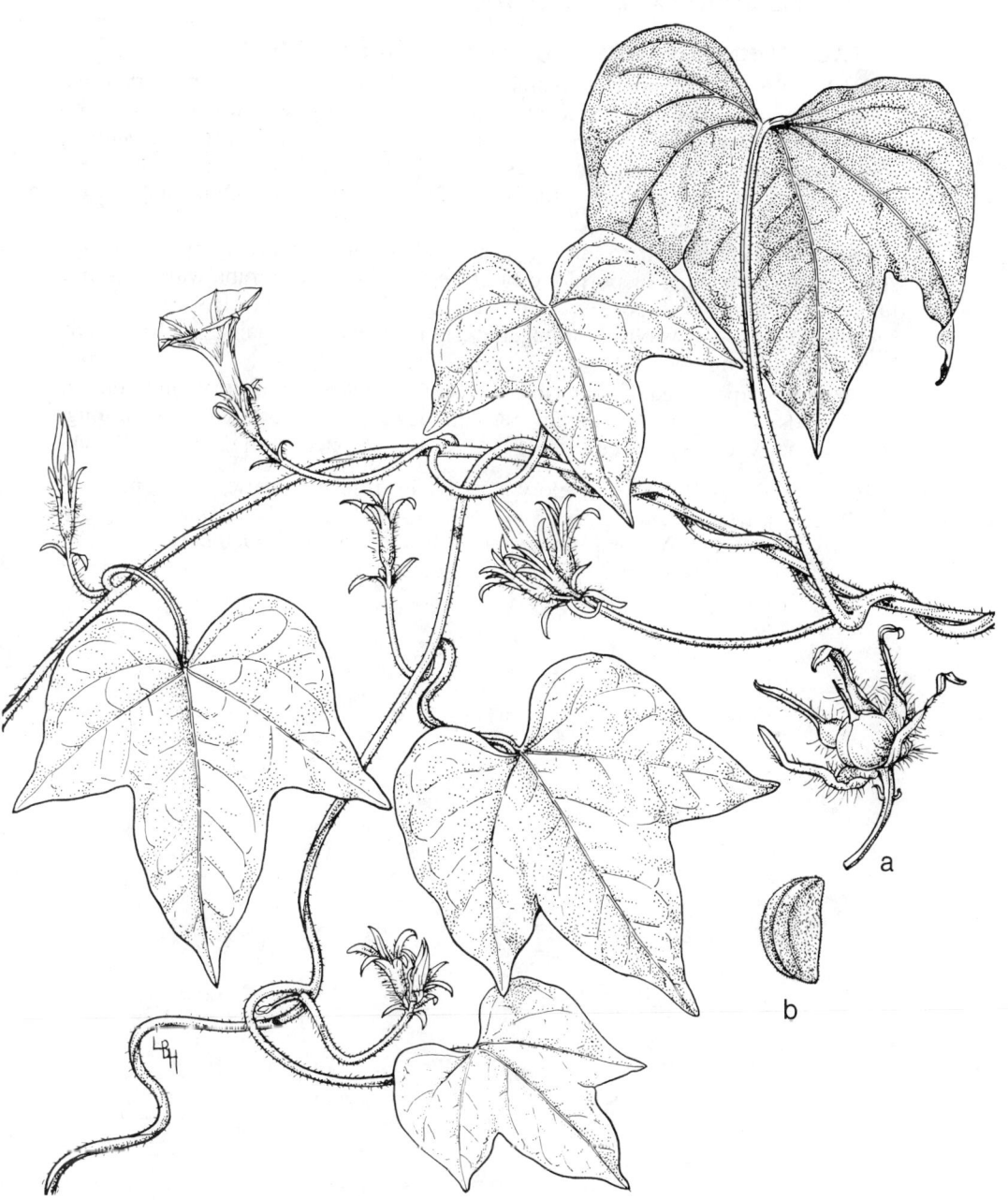

Fig. 114. Woolly morningglory *(Ipomoea hirsutula)*. Portion
of the leafy plant showing flowers and twining habit of the
stem. *a*. Seedpod surrounded by calyx lobes. *b*. Seed.

239

# TALL MORNINGGLORY

MORNING GLORY FAMILY — Convolvulaceae

### TALL MORNINGGLORY — *Ipomoea purpurea* (L.) Roth.

DESCRIPTION — A prohibited noxious weed in Arizona, tall morningglory is an annual climbing and twining vine, from a fibrous root system, which reproduces only by seeds. The twining or trailing stems are hairy, 5 to 13 feet long. Similar to woolly morningglory, but the leaves are all heartshaped and unlobed, more or less hairy, and pointed at the tip, the blades 2½ to 4 inches long, on stalks 2 to 4 inches long.

The flowers are similar to but often larger than those of woolly morningglory. They are white to blue, or purple to bright pink, with considerable variation and different markings, 1½ to 2⅝ inches long, and 1½ to 2 inches across. The 5-lobed calyx, as in woolly morningglory, is conspicuously hairy, ½ to ¾ (or rarely 1 inch) long.

The globeshaped seedpods are like those of scarlet morningglory, and contain 4 to 6 seeds. The seeds are similar, but flattened and larger, about ⅕ inch long, minutely hairy except around the scar, 3- to 4-angled, and brownish black.

DISTRIBUTION — Tall morningglory is native in tropical America. It occurs with woolly and scarlet morningglories on the farms, fields, roadsides and ditches in central and southern Arizona; flowering from about June to October.

240

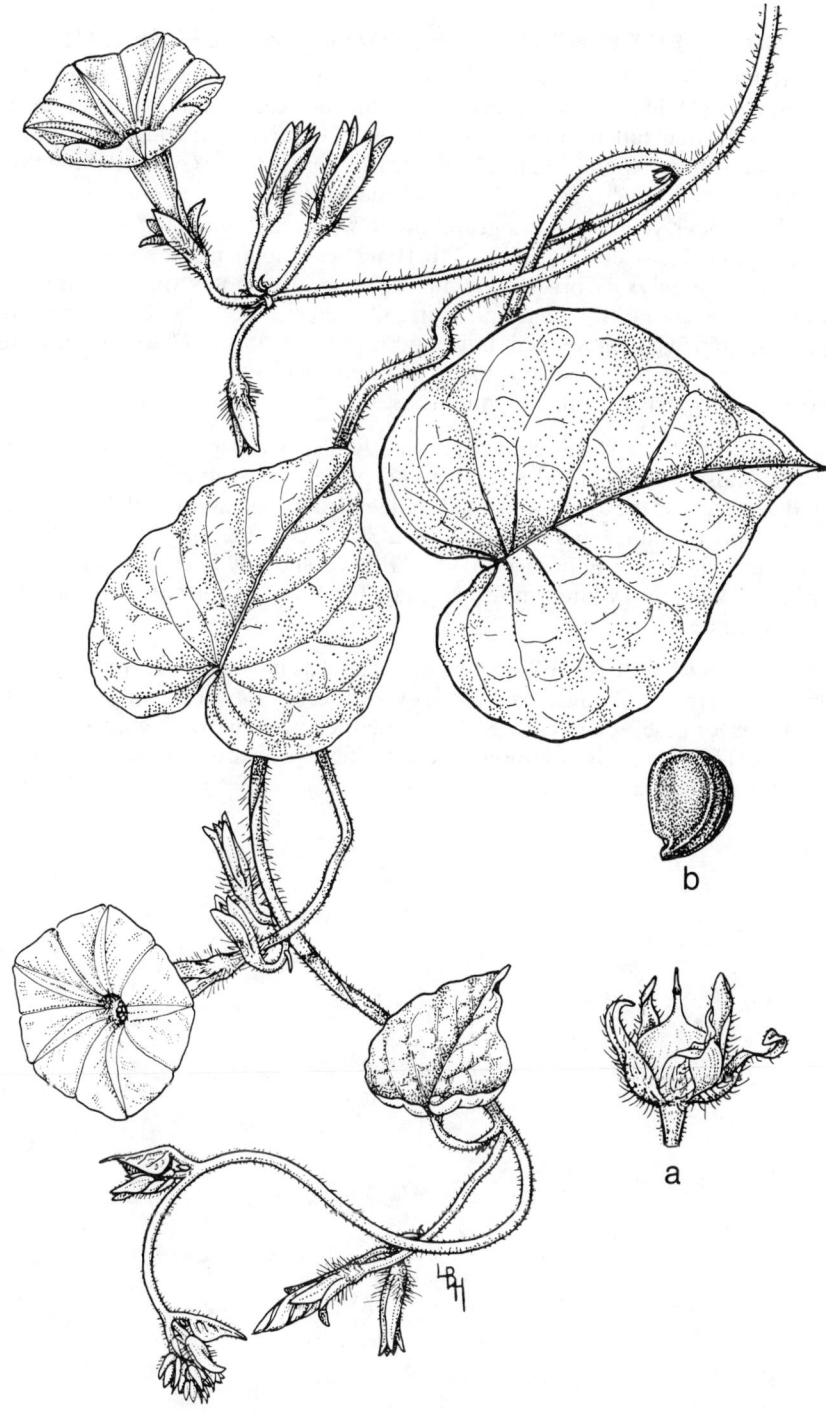

Fig. 115. Tall morningglory *(Ipomoea purpurea)*. Leafy portion of plant showing flowers and hairy, twining stems. *a.* Seedpod enclosed in calyx. *b.* Seed.

# COAST FIDDLENECK, coast buckthorn
## BORAGE FAMILY — Boraginaceae

### COAST FIDDLENECK — *Amsinckia intermedia* Fisch. & Mey.

DESCRIPTION — A conspicuously bristly yellowish annual, covered by stiff harsh hairs, reproducing by seeds only. The stems are erect, branching, ½ to 2½ feet high, and often fall on the ground in age. The alternate rough hairy leaves are lanceshaped to oblong, stalkless or nearly so, 1 to 3 (or exceptionally to 6) inches long, ⅓ to ½ (or to 1) inch broad.

The slender yellow or orange yellow flowers, ⅜ to ½ inch long, are crowded on one side of long curling spikes 3 to 10 inches long, hence the name "fiddleneck." The 5-lobed calyx is bristly, yellow, hairy, and encloses the 4 seedlike nutlets. Each nutlet contains 1 seed and is usually referred to as a seed. The nutlets are grayish, somewhat eggshaped, but 1 end is pointed, and 1/12 to ⅛ inch long. They have a sharp ridge down the back, a slightly winged angle underneath, and are covered by grayish pebblelike bumps.

DISTRIBUTION — A native weed, coast fiddleneck often is a nuisance in winter grain fields, along roadsides, and waste places. It is abundant on dry sandy or gravelly soil on the creosote desert ranges from Coconino and Mohave counties southward to Yuma and eastward to Cochise counties; mostly below 5,000 feet elevation; flowering February to May. It is relished by livestock when young, and in years of good winter moisture may form an important part of the spring forage on desert ranges.

POISONOUS PROPERTIES — The mature seeds have been demonstrated to cause hepatic cirrhosis, known as "hard liver disease" of cattle and swine, and the "walking disease" of horses. Sheep are either immune or highly resistant to the poison. The disease is common in the Pacific Northwest, but not in Arizona. This plant also may cause nitrate poisoning.

a

Fig. 116. Coast fiddleneck *(Amsinckia intermedia)*.
Bristly plant showing flowers in "fiddleneck"
inflorescence. *a*. Nutlet, containing one seed.

243

## ALKALI HELIOTROPE, Chinese pusley, quail plant
BORAGE FAMILY — Boraginaceae

### ALKALI HELIOTROPE — *Heliotropium curassavicum* L.
### var. *oculatum* (Heller) Johnst.

DESCRIPTION — A low, bluish green, fleshy perennial, reproducing by seeds and by creeping rhizomes. The plant is hairless, but covered with a whitish bloom that rubs off easily. When picked, it soon turns dark purplish brown. The stems branch from the base, are erect at first but are soon prostrate, with the tips ascending, 1 to 3 feet long. The leaves are alternate, rounded at the tip, 1 to 2½ inches long, the upper stalkless and the lower with short stalks.

The flowers are white with a yellow "eye" changing to purple, ⅛ to ¼ inch long, crowded on 1-sided, slender curling spikes which are mostly in pairs, 2 to 6 inches long. The globeshaped fruit separates into 4 seedlike nutlets when mature. The small brown nutlets, about 1/16 inch long, are beetleshaped, rounded on the back, with several raised longitudinal veins.

DISTRIBUTION — A native weed, alkali heliotrope is common in moist alkaline or saline soils, and along watercourses. Often a pest in alfalfa fields, on overflow bottom lands, irrigation ditches, canals, and roadsides, from whence it spreads along the edges of cultivated crops. Abundant in the valleys, along the banks and overflow lands of the Colorado, Little Colorado, Salt, and Gila rivers throughout Arizona; up to 6,000 feet (mostly lower) elevation; flowering throughout the year. Usually growing in dense colonies due to spreading from its extensive rhizomes.

Fig. 117. Alkali heliotrope *(Heliotropium curassavicum* var. *oculatum).*
Leafy flowering plant. *a.* Pair of inflorescences. *b.* Young fruits.
*c.* Fruit, composed of four nutlets. *d.* Single nutlet.

245

# HENBIT

MINT FAMILY — Labiatae

## HENBIT — *Lamium amplexicaule* L.

DESCRIPTION — A low slender annual, winter annual, biennial, or rarely a short-lived perennial, reproducing by seeds, by stems rooting at the lower joints, and sometimes by slender rhizomes. The weak 4-angled stems, 4 to 16 inches high, branching from the base, are erect at first but soon spreading, the lower part often reclining on the ground and rooting at some joints. The leaves are opposite, usually less than 1 inch long, hairy, and round-toothed or lobed. They are widely spaced except at the tops of the stems, often 3 to 6 inches apart. The lower ones are heartshaped on slender stalks; the upper ones much broader, stalkless and clasping the stem. The slender pink or purple flowers, ⅓ to ⅔ inch long, are tubular with 2 lips, the upper hairy on the back, and the lower spotted. The stalkless flowers are borne in the axils of the upper leaves, 6 to 10 or more forming a circle around the stem. The calyx, yellow hairy with 5 sharp bristlelike teeth, remains on the plant and encloses the 4-lobed fruit. The calyx separates into 4 seedlike nutlets, commonly referred to as seeds. They are 3-angled, 1/16 to more than 1/12 inch long, grayish brown, and sprinkled with silverish bumps.

DISTRIBUTION — Henbit is an introduced European weed, growing in moist often shady soil, but also in the sun. In Arizona it is primarily a pest in lawns, especially new lawns, also found in gardens, flowerbeds, plowed fields, and waste places. Troublesome mostly in the spring, and sometimes again in the fall at the lower elevations. Widespread throughout the southern and central part of the state and locally abundant in many areas, it is scattered in the northern part; 100 to 9,000 feet elevation; flowering February to November.

Fig. 118.  Henbit *(Lamium amplexicaule)*. Flowering plant.
*a*. Enlarged flower.  *b*. Two views of seedlike nutlets.

# HOREHOUND

MINT FAMILY — Labiatae

## HOREHOUND — *Marrubium vulgare* L.

DESCRIPTION — An upright bushy perennial which reproduces only by seeds. It has dense white woolly 4-angled stems branching from the base, ¾ to 2½ feet high. Both the stout stems and the leaves have a bitter taste. The opposite leaves are round, corrugated, and 1 to 2 inches long including the stalks. They are green above, and white woolly beneath, with rounded teeth along the margins.

The small white tubular flowers, ¼ to ⅓ inch long, are crowded into dense clusters around the stem at the base of the leaf stalks. The flowers are stalkless, and the clusters are very dense and compact around the stem. These flower groups occur at the ends of all the branches, and often extend for more than a foot on the stem. The calyx is also tubular, with 10 spinelike teeth which curve downward and are hooktipped in age. The calyx is persistent, and encloses the 4-parted fruits.

At maturity the fruit separates into 4 seedlike nutlets, which are 1-seeded and commonly referred to as seeds. They are eggshaped, brown or dark gray, about 1/12 inch long, and somewhat 3-angled; the surface has scattered dark granules.

DISTRIBUTION — Horehound is a widespread European perennial. It grows in dry soil and is a common weed of old fields, waste places, and roadsides, especially in the vicinity of permanent stock water, bedgrounds, and sheep or goat corrals. It extends throughout the state, and is abundant in many areas. It has a wide altitudinal variation, from 100 to 8,500 feet elevation; flowering May to October.

The hooked teeth of the calyx may become attached to wool or mohair, and thus lower its value. The tops of the plants are used medicinally for cough medicines, and for candy flavoring.

Fig. 119. Horehound *(Marrubium vulgare)*. Woolly plant with clusters
of flowers around the stem. *a.* Young fruit showing ten hooked spines
of the persistent calyx. *b.* Two views of seedlike nutlet.

# SMALL GROUNDCHERRY

POTATO FAMILY — Solanaceae

## SMALL GROUNDCHERRY — *Chamaesaracha coronopus* (Dunal) Gray

DESCRIPTION — Low spreading perennial which reproduces by seeds and by slender rhizomes. The plants are covered by whitish branlike flakes, which actually are flattened, branched hairs. The weak stems branching from the base are erect at first, but soon spread and recline on the ground, ½ to 1½ feet long. The alternate leaves, 1 to 4 inches long and 1/12 to ⅜ inch broad, are short stalked or stalkless. They are narrowly lanceshaped with nearly smooth margins, or more often are shallowly and irregularly lobed.

The 5-lobed flowers are greenish white or tinged with purple, wheelshaped, and ⅓ to ½ inch across. They occur singly in the leaf axils all over the plant, not just on the upper part, and are on slender stalks, which curve downward in fruit. The calyx is 5-lobed and persistent, enlarging and becoming globeshaped, but only covering about ⅔ of the mature seedpod.

The yellow berrylike seedpods, with the surrounding greenish calyx, are about ¼ inch in diameter, and hang downward. The kidneyshaped seeds, slightly more than 1/12 inch in diameter, are dark brown or reddish brown; the surface is honeycombed, roughened, and glistening.

DISTRIBUTION — Small groundcherry is a native weed, growing on dry or damp disturbed soil, in cultivated farm lands, fields, edges of walks or lawns, yards, and roadsides. Also found on mesas, bottoms, and especially on eroded or overgrazed plains on northern desert and pinyon-juniper ranges. Widespread throughout most of Arizona, but especially in the northern and central parts, where it is abundant in many areas; 2,500 to 7,500 feet elevation; flowering April to September. The Navajo and Hopi Indians are reported to eat the seedpods.

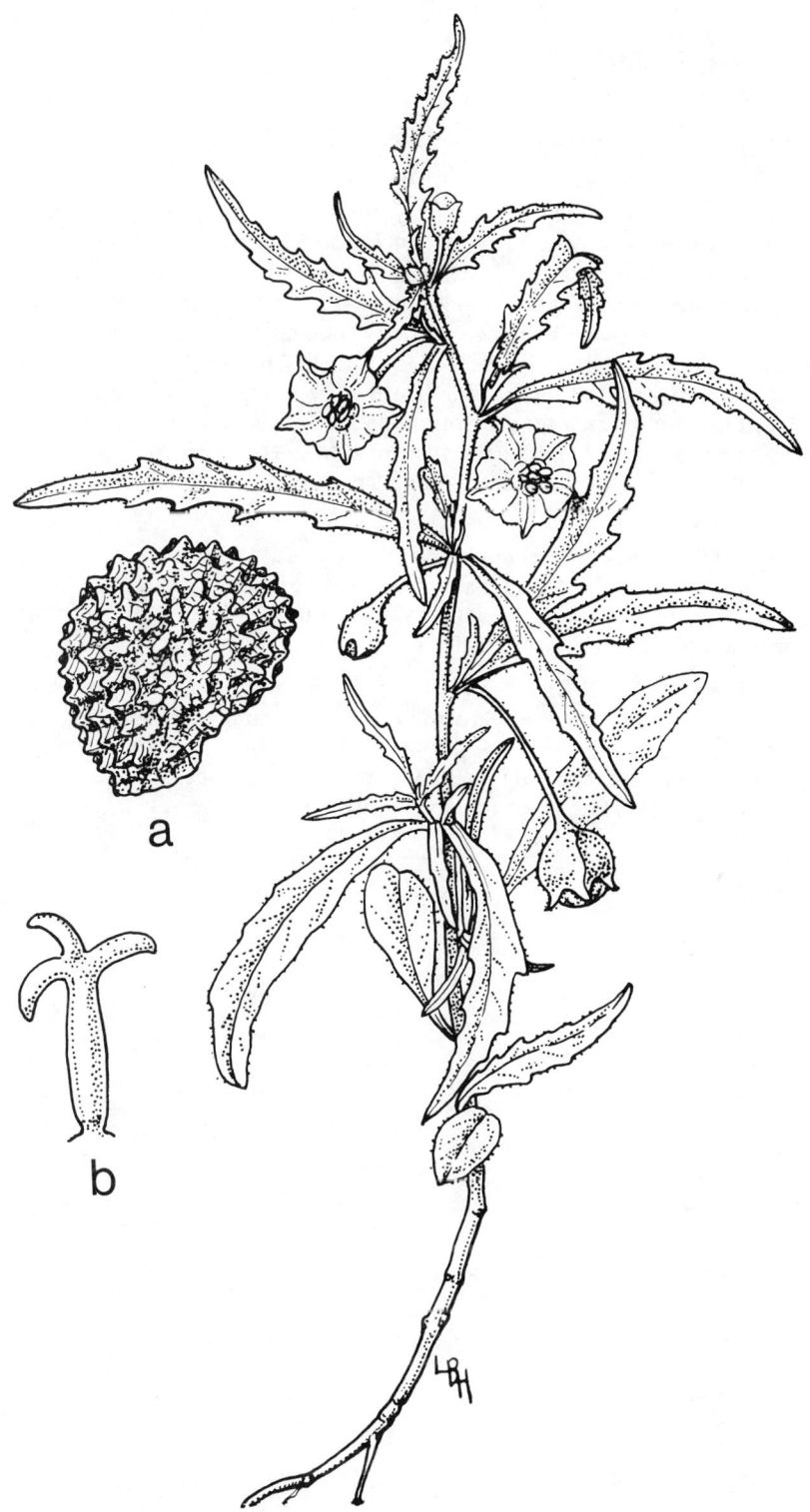

Fig. 120. Small groundcherry *(Chamaesaracha coronopus)*. Leafy plant with flowers and young seedpods. *a*. Seed. *b*. Branched hair.

# DESERT THORNAPPLE
POTATO FAMILY — Solanaceae

## DESERT THORNAPPLE — *Datura discolor* Bernh.

DESCRIPTION — An erect low annual with stout stems branching from the base, 1 to 2 feet high, which reproduces only by seeds. The plants are small and green; the leaf blades, only about 2 to 4 inches long, are slightly rounded.

The flowers are trumpetshaped, white tinged with violet, mostly 2 to 4 inches, rarely to 6 inches long, 2 inches or less across, and the margin has 10 slender teeth rather than 5. The calyx is 2¼ inches long, and 5-angled.

The large globeshaped seedpods, 1 to 1½ inches in diameter at maturity, are on thick down curved stalks. The many-seeded pods are covered by stout spines; the longer ones are ⅜ inch long when mature. The seedpods are sticky and short hairy, as are the spines. The ripe seeds are black, about ⅛ inch long, kidney-shaped, flattened; the surface is finely roughened, pitted, and bordered by a wavy grooved margin.

DISTRIBUTION — Desert thornapple is a native weed; very common in cultivated fields in some areas, as in the Yuma Valley. It is confined to southern Arizona in Cochise, Pima, Pinal, Maricopa, and Yuma counties; mostly 100 to 2,600 feet, sometimes to 4,000 feet elevation; flowering July to November.

POISONOUS PROPERTIES — All parts of the plant of the various species of *Datura* are poisonous; the seeds are the most toxic, and the young leaves next. They are poisonous to all classes of livestock and to humans. Under normal range conditions, the plants cause little trouble. Animals will not eat them unless forced to do so through starvation or confinement within heavily used pastures or corrals.

252

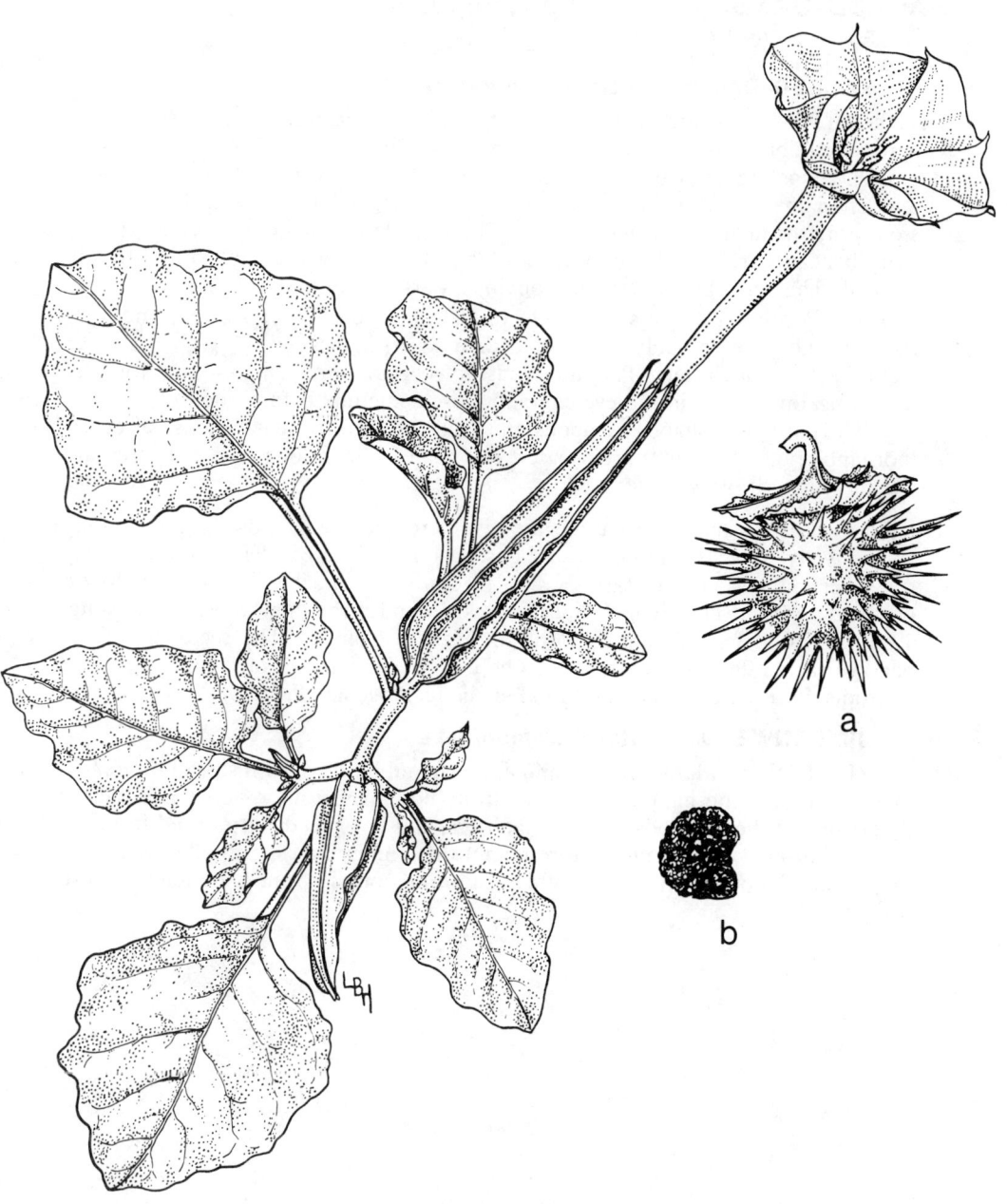

Fig. 121. Desert thornapple *(Datura discolor)*. Leafy branch
with flower and flower bud.  *a.* Spiny seedpod.  *b.* Seed.

# SACRED DATURA, Indianapple, tolguacha
POTATO FAMILY — Solanaceae

### SACRED DATURA — *Datura meteloides* DC.

DESCRIPTION — A large conspicuous grayish green perennial with a strong disagreeable odor, forming spreading clumps, reproducing by seeds only. The coarse grayish stems are erect but spreading, branched from the base, 2 to 3 feet high, and often the same in diameter. The large eggshaped leaves are alternate, on stout grayish stalks 1 to 5 inches long. The leaf blades, green above and grayish hairy beneath, are 3 to 10 more inches long, the edges wavy toothed and the tip pointed. The veins are whitish and obvious, particularly underneath.

The large showy flowers are white or pale lavender, short stalked, and very fragrant. They are broadly funnelshaped, 6 to 10 inches across, with 5 slender teeth, ½ to ¾ inch long. The numerous flowers are borne singly in the forks of the stems, open early in the evening and close sometime before noon of the next day. The hard globeshaped seedpods and seeds are very similar to those of desert thornapple, but the spines are slender, less than ⅜ inch long, and the seeds are a light yellowish brown color.

DISTRIBUTION — Sacred datura is a native perennial weed, growing in dry sandy and gravelly soils. It seldom grows densely, but is found widely scattered along roadsides, ditches, corrals, farms, waste places, and in washes, arroyos, or slopes, on the desert and in pinyon-juniper ranges. Found throughout most of the state, 1,000 to 7,000 feet elevation; flowering May to October. The seeds and other parts of sacred datura are reported to be used by Indians for medicinal purposes; the roots are used as a narcotic to induce hallucinogenic effects.

### JIMSONWEED — *Datura stramonium* L.

DESCRIPTION — A large coarse annual, green and hairless, with small flowers like those of desert thornapple. It differs from both desert thornapple and sacred datura in that the seedpods are erect, hairless, and fewspined. Introduced from the tropics, jimsonweed is common throughout the United States, but only occasional in Arizona. Known only from Cochise and Gila counties, but undoubtedly more widespread.

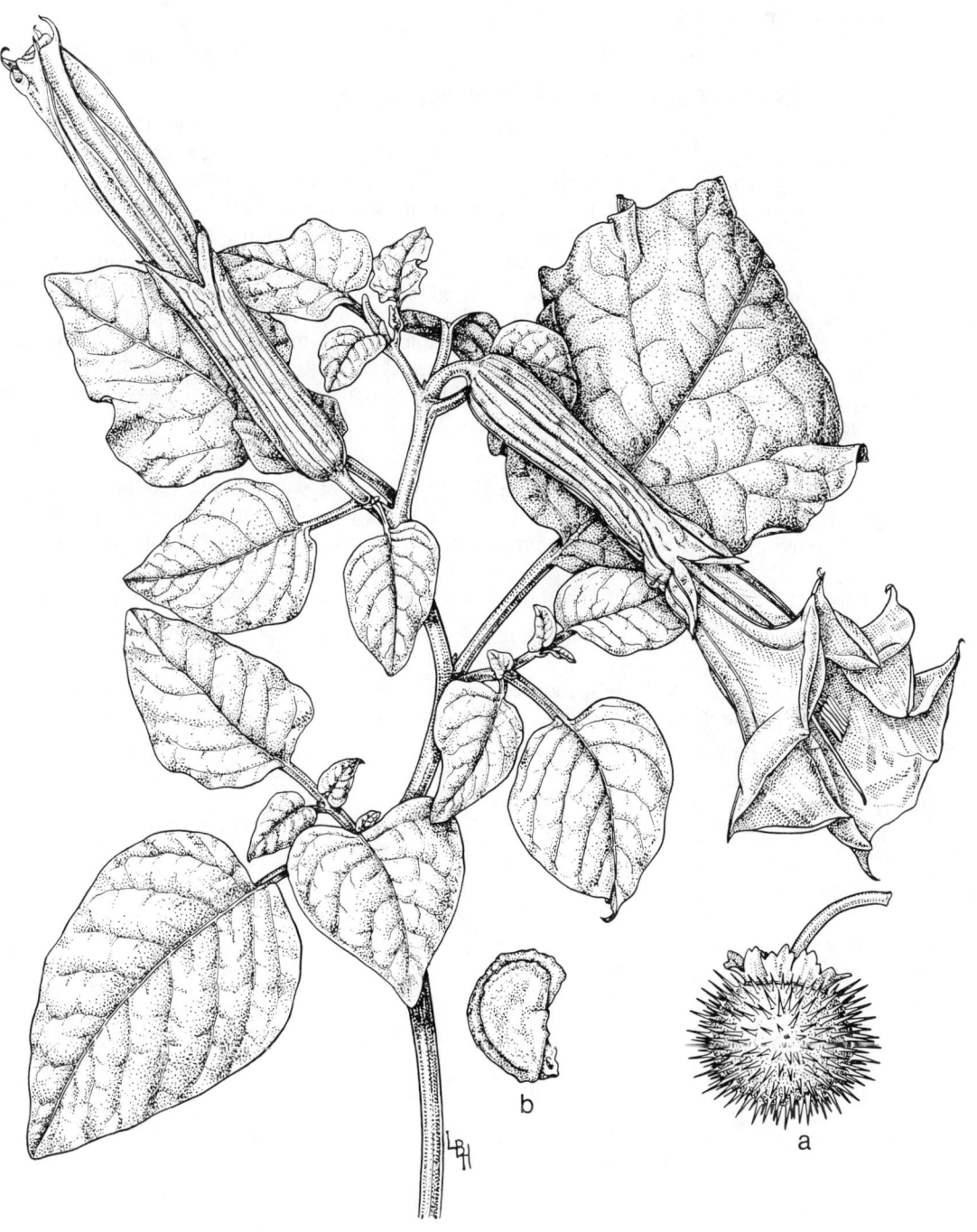

Fig. 122. Sacred datura *(Datura meteloides)*. Leafy
branch with flowers. *a.* Spiny seedpod. *b.* Seed.

# TREE TOBACCO

POTATO FAMILY — Solanaceae

## TREE TOBACCO — *Nicotiana glauca* Graham

DESCRIPTION — An evergreen, bluish green shrub or small tree, 6 to 12 (or to 21) feet high, which reproduces only by seeds. The stems are slender and loosely branching. The evergreen leaves are alternate, 1½ to 6 inches broad, ¾ to 2⅔ inches long. They are bluish green, eggshaped, and hairless, but covered with a whitish powder which rubs off easily. The margins are smooth or slightly wavy.

The long tubular flowers are yellow, about 1½ inches long, and are borne on large leafless branches at the ends of the stems. The flower tube is densely short hairy on the outside, and opens during the day. However, the 5 lobes are very short, so there is little spread. The calyx is unequally 5-toothed, and about ⅜ to ½ inch long. The seedpods are brown, many seeded, ⅜ to ½ inch long, somewhat eggshaped or oblong, on curving stalks so they hang downward. The kidneyshaped seeds are dark brown, about 1/16 inch long, with a honeycombed and roughened surface.

DISTRIBUTION — Tree tobacco is a shrubby weed of waste places, naturalized from South America. It grows in sandy or gravelly soils along roadsides, near cultivated areas, around old dwellings and ditch banks. It is common throughout the desert ranges in southern Arizona along streams, washes, and dry watercourses; 100 to 3,000 feet elevation; flowering practically the year around.

POISONOUS PROPERTIES — The leaves and young stems of tree tobacco are the most toxic parts of the plant. The plants are poisonous to all kinds of livestock and to humans. Although tobacco plants are distasteful, and where ranges provide ample forage, livestock usually do not eat them, frequent poisonings from these plants are reported.

256

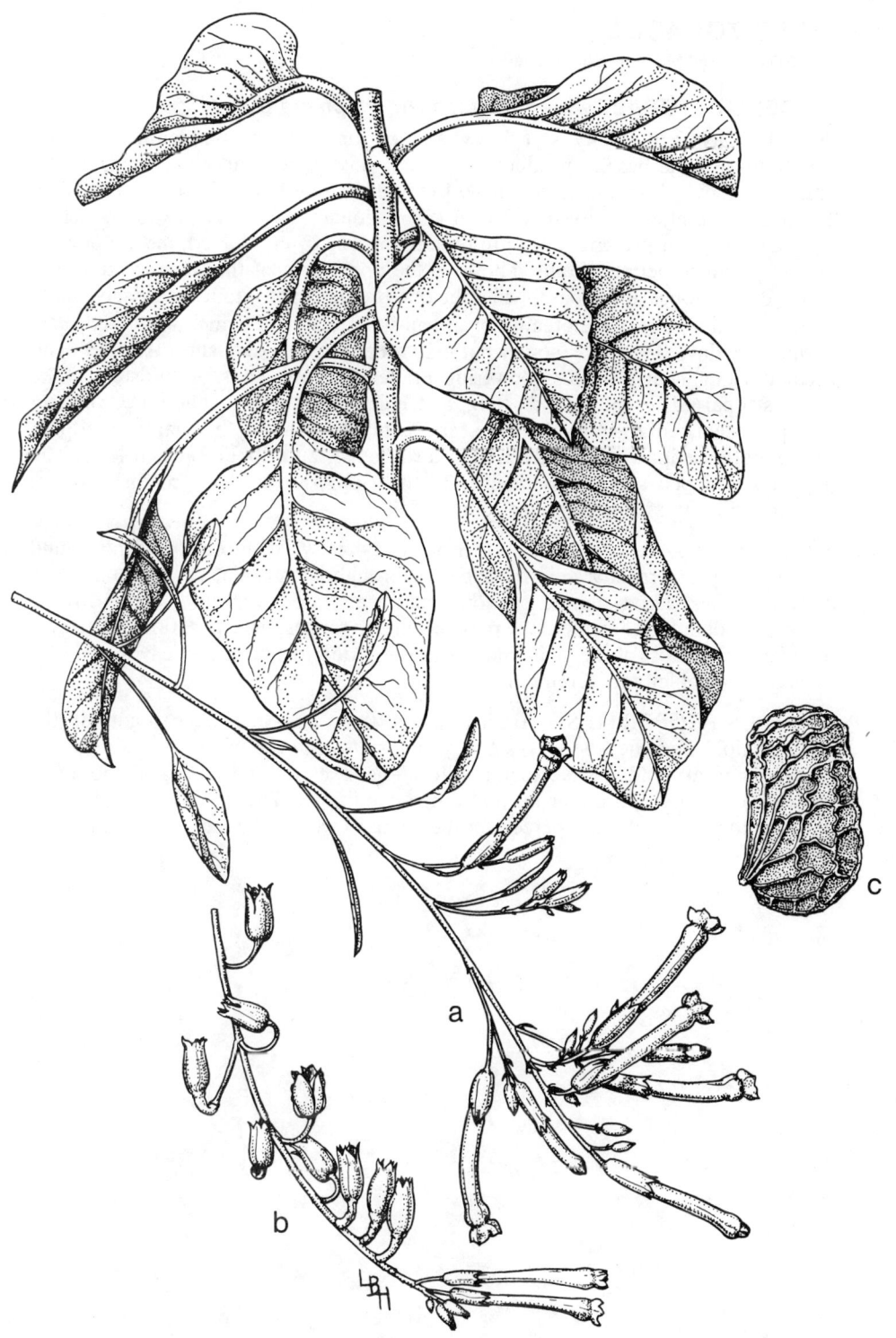

Fig. 123. Tree tobacco *(Nicotiana glauca)*. Leafy branch. *a.* Flowering branch with long tubular flowers. *b.* Fruiting branch with pendulous seedpods. *c.* Seed enlarged.

# DESERT TOBACCO

POTATO FAMILY — Solanaceae

## DESERT TOBACCO — *Nicotiana trigonophylla* Dunal

DESCRIPTION — A sticky soft hairy annual, or sometimes perennial, slightly woody toward the base, reproducing by seeds only. The stout erect stem, mostly branched above the base, is green, sticky, and densely hairy, 1 to 3½ feet high. The leaves are alternate, mostly 2 to 6 (exceptionally to 10) inches long, including the stalk when present. They are broadly oblong or eggshaped, mostly pointed at the tip, but sometimes blunt. Those on the upper part of the stem are stalkless, with 2 earlike lobes at the base, while the ones below taper into a wing-edged stalk.

The tubular flowers are creamy or greenish white, ½ to 1 inch long, and open during the day. They are somewhat hairy on the outside, short stalked, and occur on long nearly leafless flowering branches at the ends of the stems. The calyx is 6-parted, ¼ to ½ inch long, and hairy and sticky on the outer surface. It is persistent, enlarges at the base, and completely hides the mature seedpod. The smooth seedpods are many-seeded, urnshaped, and about ⅜ inch long. The tiny kidneyshaped seeds, less than 1/25 inch long, are dark or reddish brown; the surface is coarsely pitted and roughened.

DISTRIBUTION — Desert tobacco is a native range weed, most commonly found on dry sandy or gravelly soil. Widespread throughout most of the state in waste places, washes, flood plains, canyons, and water courses on desert, northern desert, woodland, and sometimes pinyon-juniper ranges; 100 to 6,000 feet elevation; flowering most of the year. Indians are reported to smoke the leaves of desert tobacco, now chiefly in ceremonials.

POISONOUS PROPERTIES — Both desert tobacco and tree tobacco contain the toxic alkaloid nicotine; the leaves and young stems are the most poisonous parts. Poisonous to all kinds of livestock and to humans; cattle and horses are poisoned more often than sheep under normal range conditions. The minimum lethal dose of desert tobacco is about 2 percent of the animal's weight, on a green weight basis.

Fig. 124.  Desert tobacco *(Nicotiana trigonophylla).*  Leafy branch with flowers and seedpods.    **259**

# WRIGHT GROUNDCHERRY
POTATO FAMILY — Solanaceae

## WRIGHT GROUNDCHERRY — *Physalis wrightii* Gray

DESCRIPTION — Stout bushy annual weeds which reproduce only by seeds. The coarse stems are mostly erect and branching from the base, sometimes spreading, 1 to 3 feet high, or 4 to 6 feet high, and 1 to 1½ inches in diameter in rich moist cotton fields where competing for light. The alternate leaves vary widely in size and shape, on stalks ¾ to 4 inches long, the leaf blades 1½ to 5 inches long, and 1/12 to 2½ inches broad. They may be lanceshaped, oblong, or egg-shaped, and mostly pointed at the tip. Usually the margins are prominently and irregularly toothed or cut, sometimes indistinctly toothed or merely wavy.

The numerous wheelshaped flowers, ½ to ¾ inch across, are whitish with a large yellow eye and purplish anthers. They are borne singly or a few together in any leaf axil, not just in the upper ones. Each is on a threadlike stalk which lengthens and curves downward in fruit. The calyx is persistent, enlarges remarkably, becoming thin and papery, ¾ to 1¼ inches long, conspicuously veiny, hangs down and looks like a green Chinese lantern.

The berrylike seedpod or "cherry," ½ to ⅔ inch in diameter, contains many seeds, and is entirely covered by the expanded calyx. The diskshaped seeds, slightly more than 1/12 inch long, are yellowish brown, with a granular surface.

DISTRIBUTION — Wright groundcherry is a native weed which has become a serious pest in the irrigated valleys of southern and central Arizona. It is abundant in all types of crops from early summer to late fall, in orchards, cotton, sorghum, vineyards, roadsides, ditches, and pastures. Also found on open range lands on sandy or gravelly soil, along streams or moist eroded slopes; 100 to 4,000 feet elevation; flowering from April to October or November.

Fig. 125. Wright groundcherry *(Physalis wrightii).* Lower portion of plant with taproot and lower leaves; flowering branch. *a.* Expanded, persistent calyx, enclosing the seedpod. *b.* Seed, greatly enlarged.

## SILVERLEAF NIGHTSHADE, white horsenettle, trompillo
POTATO FAMILY — Solanaceae

### SILVERLEAF NIGHTSHADE — *Solanum elaeagnifolium* Cav.

DESCRIPTION — A prohibited noxious weed in Arizona, silverleaf nightshade is an upright silvery perennial, usually prickly, 1 to 3 feet high, which reproduces by seeds and by deeply penetrating or creeping rhizomes. The surface of the entire plant is covered by densely matted, tiny starlike hairs, which give its characteristic silvery color.

The stems, leaves, and flower stalks may all bear slender yellowish spines; these may be scarce or sometimes wholly lacking. The thick leaves are alternate, 1 to 4 inches long (including the stalks), ¼ to 1 inch broad, and are darker above than underneath. They are lanceshaped to narrowly oblong, the margins smooth to deeply wavy.

The showy flowers are deep violet or blue, ¾ to 1 inch across, wheelshaped and 5-lobed. They are stalked, and in a few flowered clusters at the ends of the stems, or on short branches. The berrylike pods, ⅓ to ½ inch in diameter, are mottled green, dull yellow, or orange yellow when mature. They are hairless and smooth, pulpy, somewhat berrylike, and contain numerous seeds.

The nearly diskshaped seeds are about ⅛ inch long, yellowish brown, and the surface is shiny and finely granular.

DISTRIBUTION — Silverleaf nightshade is a native plant, growing preferably on moist sandy soil. It is an obnoxious weed throughout the state, but is especially troublesome in the irrigated valleys of southern Arizona, where it is a pest in all types of crops, especially cotton, sorghum, and alfalfa. Abundant on ditchbanks, row ends, along roadsides, waste places, sandy washes, and bottom lands; 100 to 5,500 elevation; flowering April to October. It is reported that the Pima Indians use the crushed berries in making cheese.

POISONOUS PROPERTIES — As little as 0.1% of the animal's weight of silverleaf nightshade has been found toxic to cattle. The ripe seedpods are slightly more toxic than the green ones, and the leaves were least poisonous. The leaves and seedpods of silverleaf nightshade contain the poisonous alkaloid solanine.

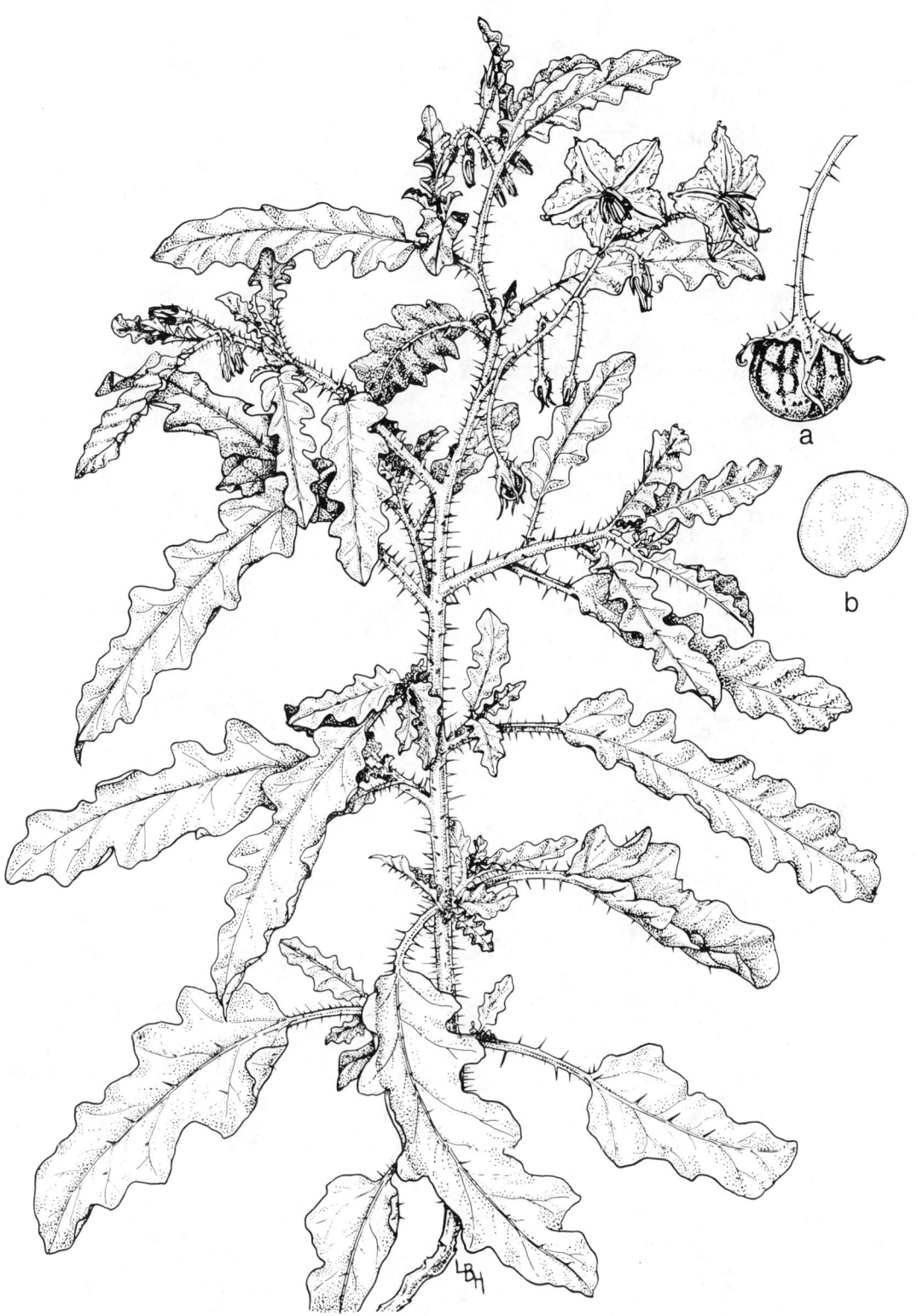

Fig. 126. Silverleaf nightshade *(Solanum elaeagnifolium)*. Spiny plant with flowers and seedpods. *a.* Mottled seedpod. *b.* Seed.

263

## BUFFALOBUR, Colorado bur, Mexican thistle, Texas thistle
POTATO FAMILY — Solanaceae

### BUFFALOBUR — *Solanum rostratum* Dunal

DESCRIPTION — A low yellow, spined, vicious soft hairy annual which reproduces by seeds. The stems, mostly branching in the upper part, are erect and bushy, ½ to 2 feet long. The entire plant, except the petals, is covered by straight yellow spines, ⅛ to ½ inch long. The leaves are alternate, 2 to 6 inches long including the stalks, irregularly cut into 5 to 7 lobes, and often these are 2- to 5-lobed. They are covered by short yellow starlike hairs, and the midribs, veins, and leaf stalks are spiny.

The yellow flowers are 5-lobed, wheelshaped, 1 to 1½ inches across, in few flowered clusters on spiny flower stalks. The calyx is covered by spines. It enlarges and forms a spiny bur, enclosing and completely covering the seedpod. The seeds are almost circular, ½ inch or slightly more in diameter, brown to reddish brown, flattened, irregularly angled, with a finely pitted surface.

DISTRIBUTION — Buffalobur, native in the Great Plains region, was probably introduced into Arizona. It grows on dry hard soil to rich moist soil of cultivated lands, as in the cotton fields of Safford (Graham County) or Avra (Pima County) valleys. Abundant throughout northern and central Arizona and increasing southward along roadsides, yards, waste places, and overgrazed plains, sometimes covering areas many miles in extent, as south of Prescott around Mayer (Yavapai County). It is obnoxious on farms or rangelands; 1,000 to 7,000 feet elevation; flowering June to September.

POISONOUS PROPERTIES — The leaves and seedpods of buffalobur contain the poisonous alkaloid solanine.

Fig. 127. Buffalobur *(Solanum rostratum)*. Portion of leafy, spiny plant
showing flowers and spiny bur, which encloses the seedpod. *a.* Seed.

# COMMON MULLEIN

FIGWORT FAMILY — Scrophulariaceae

## COMMON MULLEIN — *Verbascum thapsus* L.

DESCRIPTION — Common mullein is an erect, stout, soft woolly biennial which reproduces by seeds. The entire plant is covered by matted layers of short hairs which are forked and starlike. The large woolly stem, usually unbranched, is 2 to 6 feet high and very leafy. The leaves form a woolly rosette on the ground for the first year, from which the stem arises the second year. The stem leaves are alternate, the basal ones 6 to 18 inches long, with the upper ones gradually becoming smaller. They are crowded on the stem, nearly oblong, the tips roundish, and the upper leaves more pointed. The leaves are stalkless, but the bases are attached to the stem, and continue down it to the next leaf, thus the stem is 4-winged. The margins are smooth or slightly wavy.

The flowers are greenish yellow, stalkless, 5-lobed, and ¾ to 1 inch across. They are crowded on a long thick spike at the top of the plant, 1 to 3 feet long, and ¾ to 1¼ inches thick, sometimes with 1 to several short spikes at the base. The woolly eggshaped seedpods, about ¼ inch in diameter, contain innumerable tiny seeds. These are dark brown, less than 1/25 inch long, rodshaped with 1 end pointed, and a pitted and ridged surface.

DISTRIBUTION — Common mullein is a naturalized weed from Europe, growing in dry disturbed soil in waste places, along roadsides, railroad embankments, old dwellings, or fields. On the pinyon, juniper, and ponderosa pine ranges it is a conspicuous weed along sheep driveways, old bedgrounds, and corrals, and is of no value as forage. Widespread in northern and central Arizona from Apache to Mohave counties and abundant in many areas, southward to the Chiricahua Mountains in Cochise County; 4,500 to 8,000 feet elevation; flowering June to October.

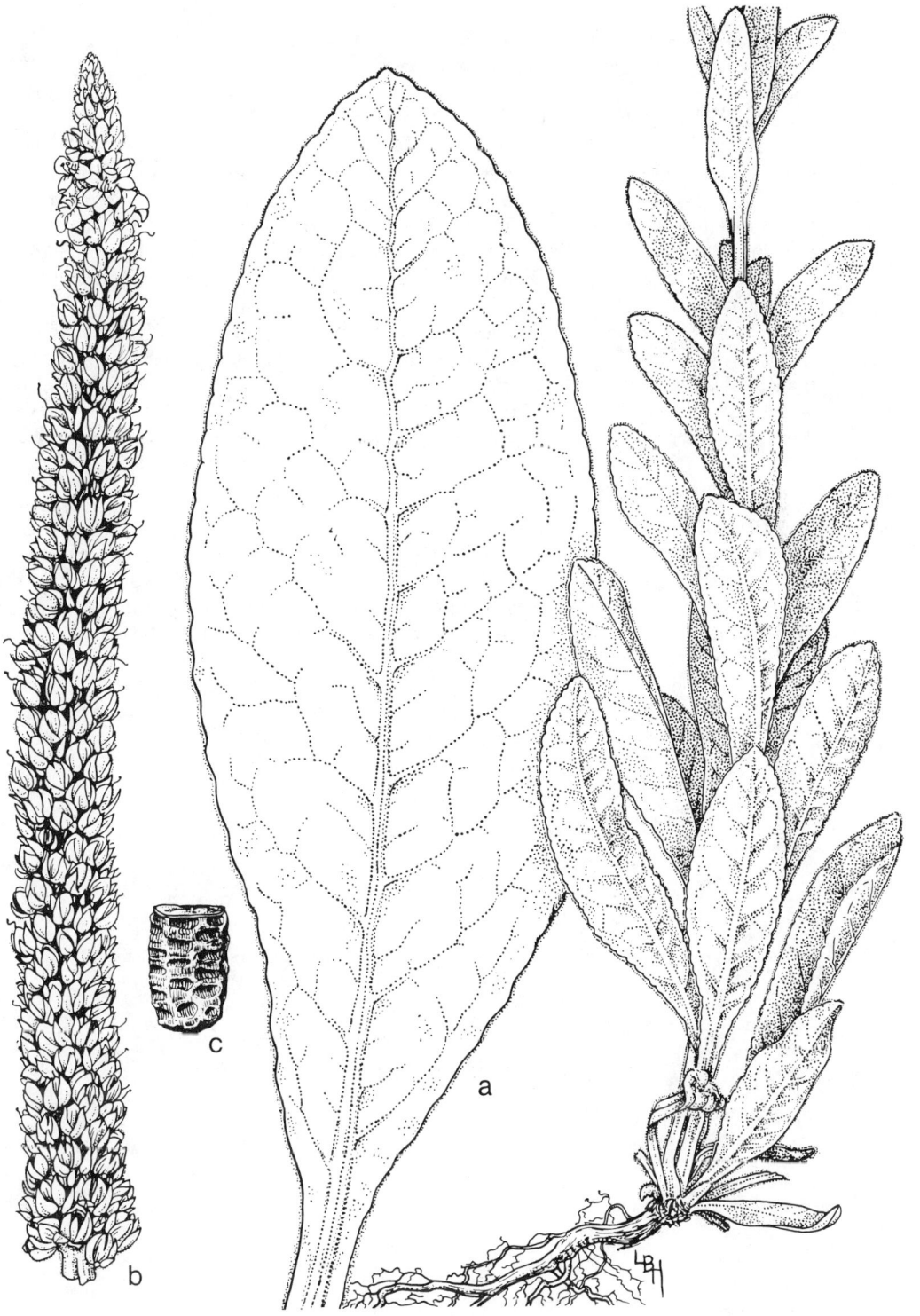

Fig. 128. Common mullein *(Verbascum thapsus)*. Lower portion of leafy plant.
*a.* Enlarged, woolly, basal leaf. *b.* Upper portion of fruiting spike, showing
flowers at the tip and seedpods below. *c.* Seed, enlarged.

267

# BUCKHORN PLANTAIN, ribwort, narrowleaf plantain
PLANTAIN FAMILY — Plantaginaceae

## BUCKHORN PLANTAIN — *Plantago lanceolata* L.

DESCRIPTION — A low erect perennial, 1 to 1½ feet high, with narrow dark green leaves, from a thick fibrous root system, which reproduces by seeds, and sometimes by new shoots from the roots. The leaves are 3 to 12 inches long (including the stalk), ¼ to 1 inch broad with smooth, wavy, or barely toothed margins. They are strongly 3 to 5 ribbed, oblong or lanceshaped, tapering at the base into a slender stalk. The leaf axils are often filled with long brownish cottony hairs.

The flowers are similar to those of broadleaf plantain, but occur in short thick spikes ¾ to 2 inches long at the tip of the flower stalks, which are much longer than the leaves. (In plantain the flower stalks are rarely longer than the leaves.) The seedpods are globeshaped, dry and papery, about ⅛ inch long, contain 2 seeds, and open by the upper half falling off as a lid. The seeds are 1/16 to 1/12 inch or slightly more long, boatshaped, the surface usually shiny, and greenish brown to dark brown.

DISTRIBUTION — Buckhorn plantain, a European introduction, is primarily a pest throughout most of Arizona. It also grows in moist soil of fields, dooryards, waste places, irrigated pastures, gardens, and along streams; 100 to 8,000 feet elevation. More troublesome in southern Arizona irrigated fields than broadleaf plantain, it is an annoying pest in alfalfa, small grains, lawns, pastures, roadsides, and waste places; flowering from April to October.

Fig. 129. Buckhorn plantain *(Plantago lanceolata)*.  Leafy plant with short
flowering and fruiting spikes at tip of the long flower stalks.  *a*. Two views of seed.

# BROADLEAF PLANTAIN, common plantain, rippleseed plantain
PLANTAIN FAMILY — Plantaginaceae

## BROADLEAF PLANTAIN — *Plantago major* L.

DESCRIPTION — Low tufted perennial from a thick fibrous root system which reproduces by seeds and sometimes by new shoots from the roots. There are no true stems above the ground; those bearing the flower spikes are flower stalks. The large dark green hairless leaves are all at the base of the plant. The leaves are broadly eggshaped or oval, 3 to 8 inches long, on stalks 2 to 5 inches long and 2 to 4 inches broad, prominently 5- to 7-ribbed, with smooth or usually wavy to toothed margins.

The numerous whitish or colorless flowers are small, 4-lobed, thin, dry, and persistent. They are crowded along a narrow elongated spike, 3 to 12 inches long and ¼ to ⅓ inch thick, on the upper part of the slender leafless flower stalk. The seedpods, similar to those of buckhorn plantain, contain 6 to 20 reddish brown seeds. The seeds are 1/25 to 1/16 inch long, and somewhat angled. The surface is granular, with fine radiating lines.

DISTRIBUTION — Broadleaf plantain is a naturalized weed from Europe, and primarily a pest in lawns throughout most of Arizona. It also grows in the moist soil of fields, dooryards, waste places, irrigated pastures, gardens, and along streams; 100 to 8,000 feet elevation; flowering March to October. Its tufted growth habit, large coarse leaves, and long flowering spikes are unsightly in lawns, and thus it is a particularly objectionable weed.

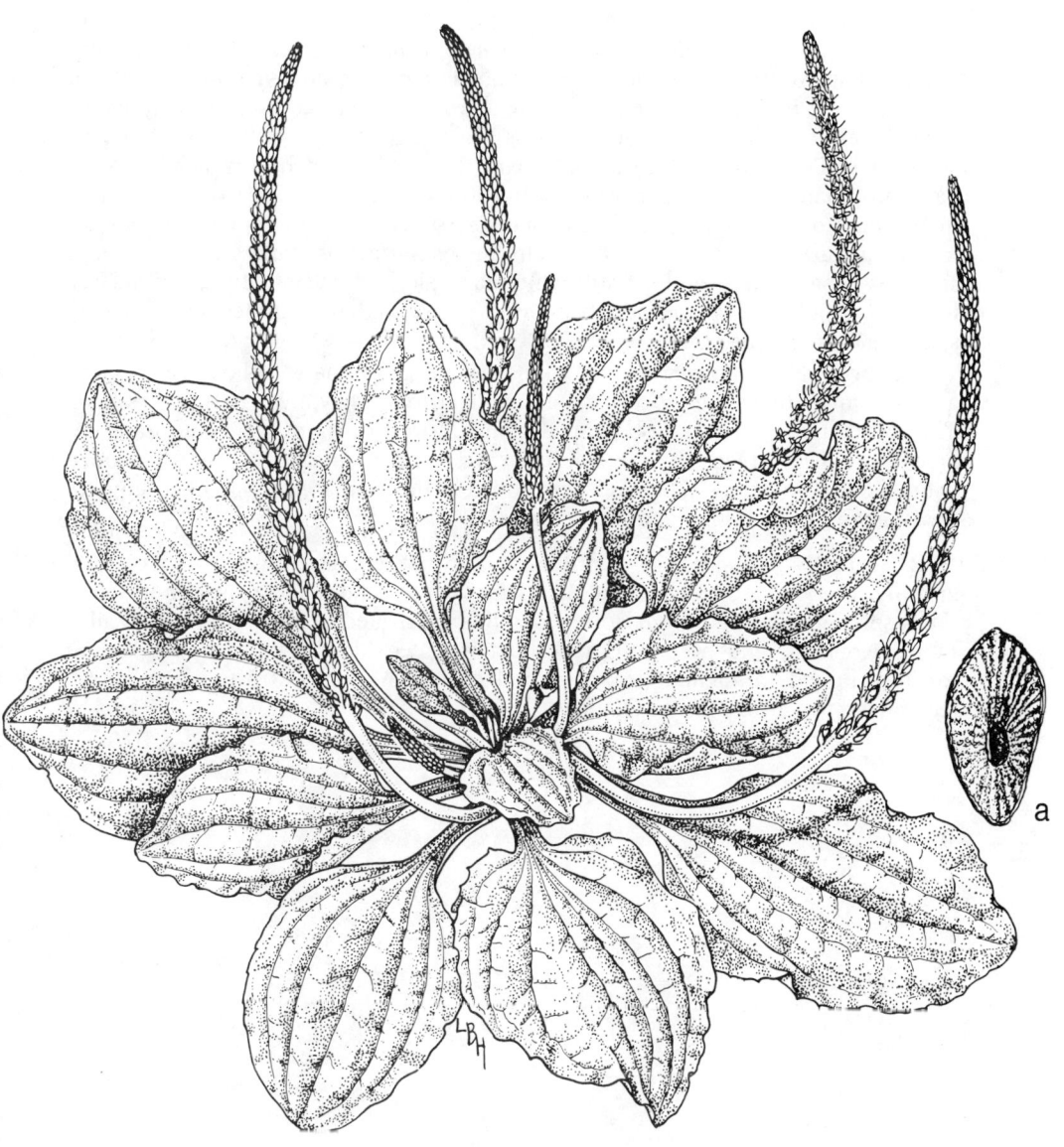

Fig. 130. Broadleaf plantain *(Plantago major)*. Leafy plant
with elongated flowering and fruiting spikes. *a.* Seed.

271

## FINGERLEAF GOURD, fingerleaf coyote melon
GOURD FAMILY — Cucurbitaceae

### FINGERLEAF GOURD — *Cucurbita digitata* Gray

DESCRIPTION — Coarse perennial trailing or climbing vine, from a long, deep-seated, fleshy tuberlike taproot several inches in diameter, reproducing by seeds and by stems rooting at the joints. The tough stems are angled 3 to 10 feet (or to 40 feet in moist cotton fields) long, 1 to a few from the root, branching immediately and radiating from the root. The leaves are widely spaced on the stems, 3 inches to more than a foot apart. They are divided very nearly to the base into 5 lanceshaped narrow lobes, 1½ to 10 inches long, ⅛ to more than an inch broad, the margins smooth, toothed, or sometimes with 1 or more pairs of lobes, on stalks 1 to 5 inches long. The upper surface is dark green with conspicuous broad bands of short stiff white hairs along the veins, the lower surface uniformly rough hairy, varying from very dense and the surface gray to very sparse and the surface green.

The large yellow flowers are of 2 kinds, male and female, although they have a similar appearance and occur in the same leaf axil. They are 2 to 4 inches long. The seedpod is a smooth globeshaped gourd, 2 to 3 or more inches in diameter, green with pale green longitudinal markings, and at maturity yellowish green with pale yellow stripes. The seeds embedded in the pulp are shaped like pumpkin seeds, about ⅜ inch long and tan colored.

DISTRIBUTION — Fingerleaf gourd is a native weed, usually growing on dry, gravelly, or sandy soil of roadsides, waste places, plains, and mesas, but sometimes spreading into adjacent cotton or other cultivated fields as in Avra Valley (Pima County), and becoming a pest on irrigated lands. Often a nuisance on ranches. Found throughout southern and central Arizona; 100 to about 5,000 feet or less; flowering June to October.

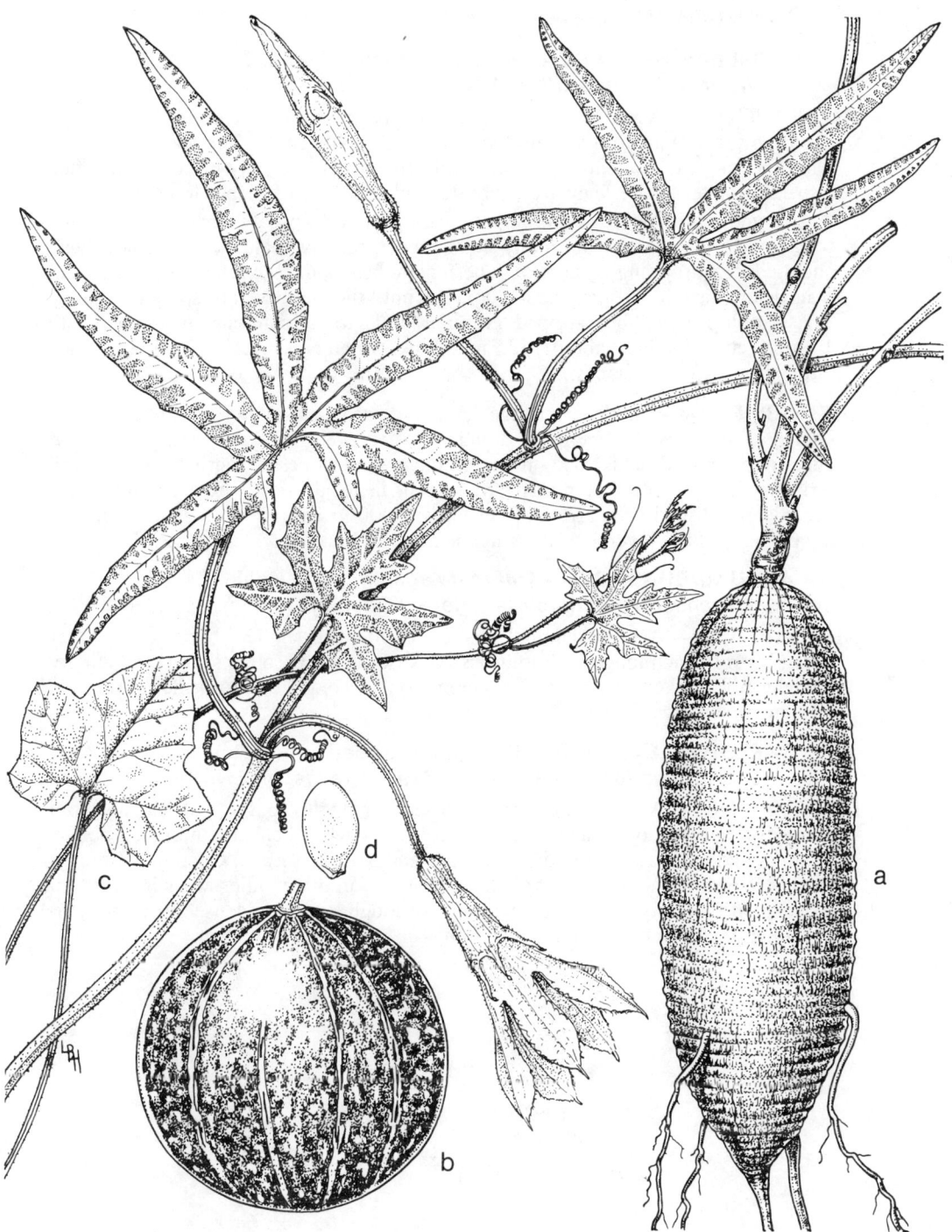

Fig. 131. Fingerleaf gourd *(Cucurbita digitata)*. Portion of vine showing tendrils, five-lobed, fingerlike leaves, and flowers. *a.* Fleshy tuberlike taproot. *b.* Gourd (fruit). *c.* Leaf of young plant (seedling). *d.* Seed, enlarged.

## SLIMLEAF BURSAGE, bursage ragweed
SUNFLOWER FAMILY — Compositae

### SLIMLEAF BURSAGE — *Ambrosia confertiflora* DC.
### (*Franseria confertiflora* [DC.] Rydb.)

DESCRIPTION — A very leafy, somewhat bushy perennial, 1 to 3 feet high, reproducing by seed and proliferating by slender, creeping roots. The leaves, often so hairy they appear gray, are divided into narrow lobes that may be further divided into still smaller lobes. They are alternate, and about 2 to 5 inches long.

The many narrow flowering stems, arising at the tips of the branches, are 2 to 5 inches long, and bear many small heads of tiny male flowers, each head enclosed in a drooping green cup. The female heads, densely clustered in the leaf axils below the male heads, are seldom seen until they mature into spiny little burs. These are topshaped, beak-tipped, granular, 1/12 to ⅛ inch long, and armed with 10 to 20 curved spines, about 1/25 inch long, which end in a definite hook. Each bur encloses 1 or 2 achenes.

DISTRIBUTION — Slimleaf bursage is native, growing in dry or moist, rocky, sterile, or fertile soil. Abundant along city streets, highways, waste places, and edges of cultivated fields, also on barren mesas and slopes in southern and central Arizona; less troublesome northward, except in local areas where colonies have become established (e.g. around Flagstaff); 1,000 to 7,000 feet elevation. Flowering from April to November, its pollen is not a serious hayfever hazard.

### ANNUAL BURSAGE — *Ambrosia acanthicarpa* Hook.
### (*Franseria acanthicarpa* [Hook.] Coville)

DESCRIPTION — A bushy annual, distinguished from slimleaf bursage by larger burs (⅛ to ⅓ inch long) with longer spines (1/12 to 3/16 inch long), which are straight, not hooked at the tip. Very common in sandy soil in northern Arizona, infrequent southward.

### SKELETONLEAF BURSAGE, Bur ragweed —
### *Ambrosia tomentosa* Nutt.   (*Franseria discolor* Nutt.)

DESCRIPTION — A perennial similar in growth habits to slimleaf bursage, but easily distinguished by its leaves, which are silvery white beneath and green above. In yards and cultivated ground at Flagstaff and Mormon Lake, Coconino County, local in Winslow (Navajo County), and probably in many other areas in northern Arizona; at 5,000 to 7,500 feet elevation. Although known to cause nitrate poisoning in livestock, the extent or losses (if any) in Arizona is unknown.

Fig. 132. *A.* Slimleaf bursage *(Ambrosia confertiflora).* Branch showing
finely-divided leaves, with separate male and female flower heads, the
male heads at the top and the female heads in the leaf axils beneath
them. *Aa.* Enlarged male head with male flowers. *Ab.* Mature female
head (bur) with curved, hooked spines, enclosing 1 (or 2) achenes.
*B. (A. acanthicarpa).* Mature female head (bur) with straight spines.

275

# SPINY ASTER, Mexican devilweed
SUNFLOWER FAMILY — Compositae

## SPINY ASTER — *Aster spinosus* Benth.

DESCRIPTION — Grayish, much branched perennial, nearly leafless, 2 to 9 feet high, reproducing by seeds and by widespreading creeping rhizomes. The intricately branched grayish stems are hairless, longitudinally ribbed, woody below, with few to many very stout, stiff, sharp, greenish spines, ⅛ to 1 inch long, borne in or above the axils of the upper leaves.

The upper leaves are scarce, scalelike but green, and soon fall off. The lower leaves, also scarce, are very slender, ¼ to 2½ inches long. The flower heads are ⅓ to ½ inch high, and ½ to ¾ inch across, including the many narrow white "petals" (ray flowers); the center of the head is yellow or brownish.

The flower heads are many to numerous, stalked, and borne at the top of the plant on branching flowering stems, or sometimes they are solitary at the end of a branchlet. The reddish brown achenes, 1/16 to 1/12 inch long, are narrowly oblong, hairless, longitudinally ribbed, with a tuft of fine hairs at 1 end.

DISTRIBUTION — Spiny aster is a native plant, commonly forming rank, hedgelike thickets along the banks of irrigation ditches. In some agricultural areas, it infests cultivated crops, especially soybean, cotton, alfalfa or small grain fields, as in the Yuma Valley. Also a common pest in heavy, more or less alkaline valley soils, or in moist saline soil along river bottoms, pastures, and low places. Found throughout the state in the deserts; 100 to about 4,500 feet elevation; flowering April to October, but principally in the summer and fall.

Fig. 133.  Spiny aster. *(Aster spinosus).*  Flowering branch; also,
enlarged stem showing spines.  *a.* Achenes with tuft of  hairs at top.

## SEEPWILLOW BACCHARIS, seepwillow, waterwally
SUNFLOWER FAMILY — Compositae

### SEEPWILLOW BACCHARIS — *Baccharis glutinosa* Pers.

DESCRIPTION — Straggling shrubs with willowlike leaves and several clustered stems 3 to 10 (or 13) feet high, reproducing only by seeds. The stems are woody below, ½ to 1 inch diameter at base, with green, prominently grooved branches. The bark on the old stems is dark gray and furrowed. The numerous leaves are alternate, bright green, leathery, lustrous from sticky resins, and have a rather pleasant odor, narrowly lanceshaped, 1½ to 6 inches long, ¼ to ⅔ inch broad, and with smooth margins and small but definite teeth.

The small, unattractive, greenish yellow flowering heads, about ⅕ inch diameter, occur in dense clusters at the tips of the main branches only, and are not scattered in the leaf axils. Male and female heads grow on separate plants. These plants are clearly distinguishable at the flowering and fruiting stages. The narrowly oblong seedlike achenes, 1/25 to 1/16 inch long, are strawcolored to greenish brown, with 5 white longitudinal ribs, and at one end bearing a thin tuft of fine, silky whitish hairs, ⅕ inch long.

DISTRIBUTION — A native shrub of watercourses, semipermanent irrigation canals or ditches, and streambanks. Very common along the Colorado River drainage from Apache to Mohave counties, and river bottoms, disturbed areas, and depressions where water collects; sometimes in moist soil at ends of cotton rows; throughout southern Arizona in the desert and desert grassland ranges; from 100 to 5,700 feet elevation; flowers from March to December.

Seepwillow baccharis is unpalatable, and worthless as forage. It often forms dense thickets along streambeds, and in water-conscious Arizona is considered to make excessive use of precious water. It also may clog stream channels, and cause flash floods to back up and inundate adjacent lands.

Because of its rapid growth and deep fibrous roots, it has been used extensively in erosion control plantings along watercourses, and is propagated from cuttings. It is interesting to note that seepwillow baccharis, like tamarix, is now considered an objectionable weed due to characteristics which were considered virtues at one time.

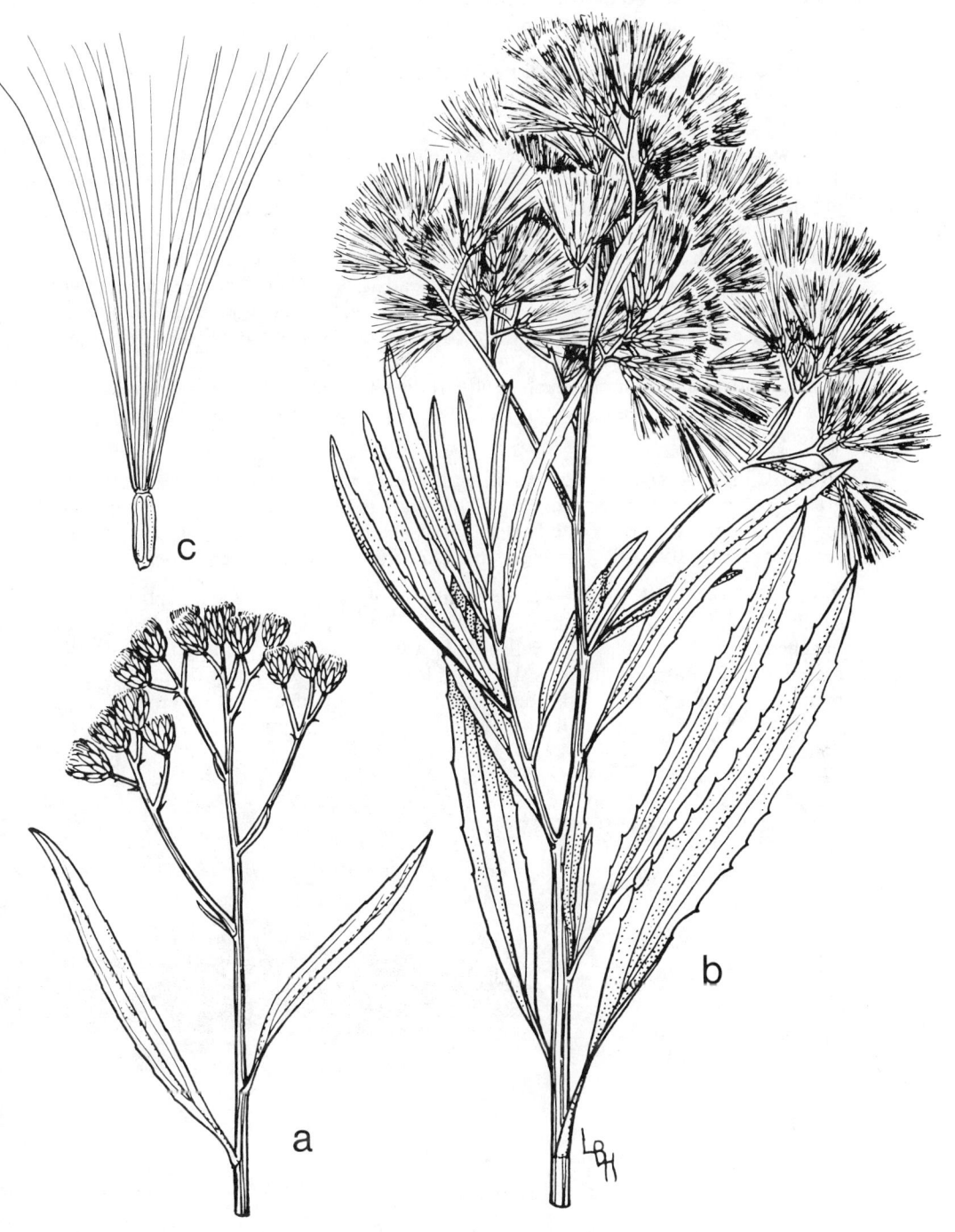

Fig. 134. Seepwillow baccharis *(Baccharis glutinosa)*. *a*. Branch showing
male heads, from male plant. *b*. Leafy branch from female plant with female
flower heads in fruit. *c*. Seedlike achene with tuft of fine silky hairs.

279

## DESERT BAILEYA, desert marigold
SUNFLOWER FAMILY — Compositae

### DESERT BAILEYA — *Baileya multiradiata* Harv. & Gray

DESCRIPTION — Low, white woolly, tufted annual or often perennial from a thick, almost woody taproot, reproducing only by seeds. There are several to numerous erect woolly stems, eventually much branched from the base, ½ to 1½ feet high. They are leafy only at the base or below the middle. At the top, each stem becomes the stalk for 1 head with many flowers. The soft, woolly leaves are alternate, clustered largely at the base of the plant and divided into irregular close set lobes, or some are divided again. The upper leaves are few lobed, or may be narrow and smooth edged. The showy yellow flower heads are 1 to 1¾ inches across, with 20 to 50 bright yellow ray flowers ("petals"). These do not drop off, but are persistent and become faded, papery, and turned downward in age. The flower heads are mostly borne singly at the tips of the stems and branches on long naked stalks, 4 to 8 inches long. Each head produces at least 100 achenes, which are rodshaped, about ⅛ inch long, light brown, with many longitudinal nerves on the surface.

DISTRIBUTION — Desert baileya is a native range weed growing on sandy or gravelly soils. A most attractive and abundant plant along roadsides, plains, and mesas throughout most of the state, adding beauty to the desert and desert grassland ranges; up to 5,000 feet elevation; flowering March to November.

POISONOUS PROPERTIES — Desert baileya, either fresh or dried, is poisonous to sheep and goats, but not to horses or cattle. The plant is not palatable to sheep, but the showy flower heads are relished. However, the flowering and fruiting heads are nearly twice as poisonous as the green leaves. Goats evidently do not graze the plant under range conditions, but have been poisoned in experimental feeding. Sheep losses from desert baileya have occurred in Arizona when green forage is scarce.

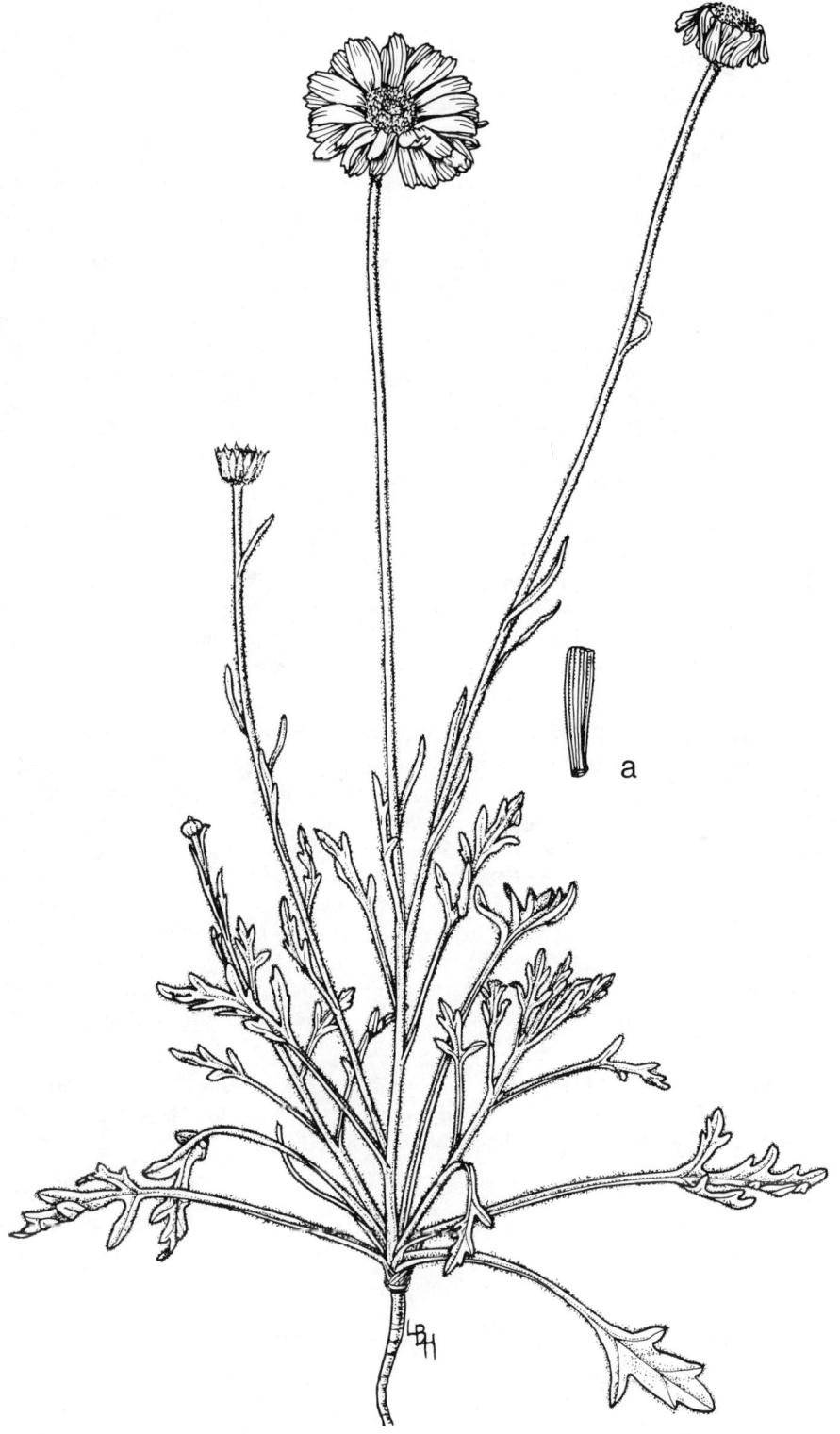

Fig. 135. Desert baileya *(Baileya multiradiata)*. Woolly plant with showy flower heads on long stalks, the rays persistent and pendulous in age. *a.* Achene longitudinally ridged.

281

# MALTA STARTHISTLE, tocalote
SUNFLOWER FAMILY — Compositae

### MALTA STARTHISTLE — *Centaurea melitensis* L.

DESCRIPTION — Grayish green annual from a taproot, with winged stems, 1 to 2
feet high, branched mostly above the base, reproducing by seed. The stem leaves
and alternate, narrow, and unlobed, with the bases prolonged down the ridged
stems as narrow green wings. The basal rosette leaves are 2 to 5 inches long, and
cut into several pairs of lobes, with the end lobe much larger.

Each flower head, bearing small tubular yellow flowers, is about ½ inch high,
and woolly in bud. Each head is enclosed by stiff bracts which bear slender spines
⅛ to ⅜ inch long, usually purplish at the base. The largest spine of each of the
middle bracts is usually branched below, with smaller spines at the base. The achene
is smooth, grayish to dark brown, somewhat longitudinally striped, and about ⅛
inch long with a hooklike notch on one side just above the base. At the top it bears
about 3 rows of whitish bristles of unequal length.

DISTRIBUTION — Malta starthistle, originally introduced into the United States
from Europe, is occasional in alfalfa fields, grain fields, pastures, the banks of
irrigation ditches, roadsides, and waste places, in Apache, Yavapai, Maricopa,
Pinal, Graham, Pima, and Cochise counties; mostly below 4,000 feet, but sometimes to 7,000 feet elevation; flowering May to July or spasmodically to frost. Not
abundant enough in any state (except Oregon) to be so serious as to be declared
a noxious weed.

### YELLOW STARTHISTLE — *Centaurea solstitialis* L.

DESCRIPTION — Annual with yellow flowers and winged stems resembling Malta
starthistle, but differing in that the spines on the middle bracts enclosing the heads
are stout, bright yellow, unbranched, and the longest is ½ to 1 inch long. The
achenes are also notched on one side just above the base, but the notch is shallow
and not clearly hooklike. On the outermost achenes, the bristles at the top are
lacking.

Yellow starthistle, another European introduction, is not declared a noxious
weed in Arizona, but it is in seven other western states, where it is particularly
troublesome in small-grain and alfalfa fields. Uncommon in Arizona; occasional
in Coconino, Mohave, Yuma, and Pima counties, and probably elsewhere in
Arizona.

POISONOUS PROPERTIES — In California, horses on ranges where yellow starthistle is common and green forage is scarce may develop a nervous disorder
known as "chewing disease." It is unknown in Arizona.

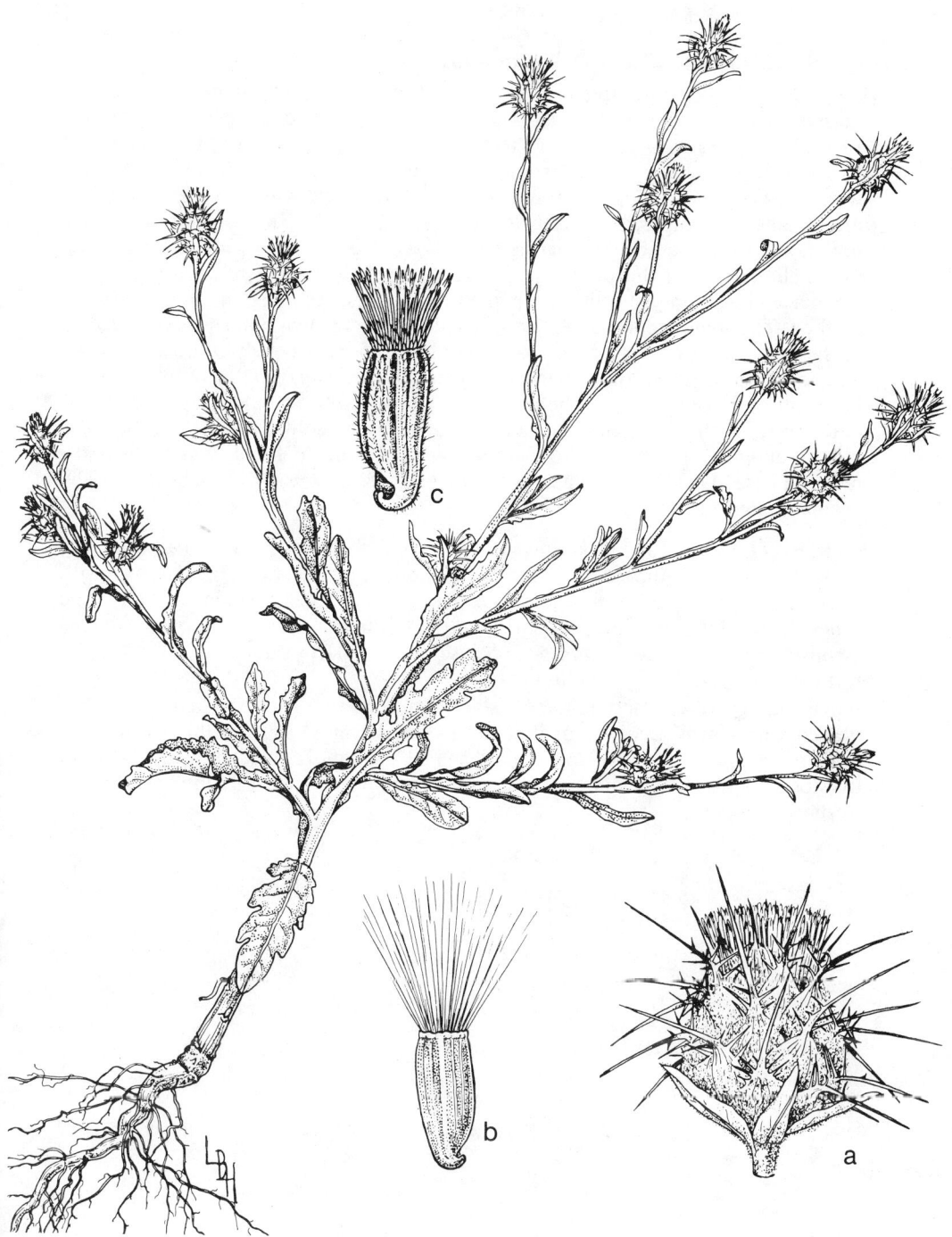

Fig. 136. Malta starthistle *(Centaurea melitensis)*. Leafy plant with flower heads. *a.* Flower head, enclosed in spiny bracts, the spines of the middle bracts branched below. *b., c.* Achenes with basal hooklike notch on one side.

## RUSSIAN KNAPWEED, turkestan thistle
SUNFLOWER FAMILY — Compositae

### RUSSIAN KNAPWEED — *Centaurea repens* L. (*C. picris* Pall.)

DESCRIPTION — A prohibited noxious weed in Arizona, Russian knapweed is a
bushy, many-branched perennial, 1 to 3 feet high, from creeping horizontal and
vertical underground stems, 2 to 4 feet deep, the older ones dark brown to black.
Reproducing by leafy shoots from the underground stems and by seeds. The
wingless, very leafy stem is covered with soft gray hairs when young. The leaves
are of different shapes, the basal deeply lobed, 2 to 4 inches long, ½ to 1 inch
broad, and form a quickly-withering rosette on the ground. The lower stem leaves
are smaller, lobed or sharply toothed, and the upper leaves, ½ to 1½ inches long,
are narrowly oblong, the tip sharp pointed and the margins smooth or slightly
toothed. The small, coneshaped flower heads, ¼ to ½ inch diameter, and solitary
at the tip of leafy branchlets, are composed of about 16 deeply 5-lobed, tubular,
lilac to bluish to rose pink flowers. Surrounding the flower heads are many pearly
bracts in overlapping series, the outer series broad with rounded papery tips, the
inner series with very hairy long taillike tips. The seedlike achenes are oblong,
plump, about ⅛ to ¼ inch long, and grayish or ivory colored, without a notch
near the base. These bear numerous whitish bristles on the top, which drop off by
achene maturity.

DISTRIBUTION — Russian knapweed is a noxious native of eastern Europe and
Asia. Grows in disturbed soil, forming colonies in cultivated fields, orchards,
ditches, pastures, roadsides, and waste places. Sometimes abundant locally, and
a pest in sorghum, alfalfa, and small grains. Scattered from Navajo to Yavapai
counties. Southward to Pima and Cochise counties; from about 1,000 to 7,100
feet elevation; flowering May to October.

Russian knapweed should be considered a very serious weed, and its spread
viewed with alarm. Because of its efficient underground system, this weed, once
established, is almost impossible to eradicate. This plant is avoided by all classes of
livestock because of its very bitter taste. Infestations have steadily increased in the
western states.

284

Fig. 137. Russian knapweed *(Centaurea repens)*. Lower part of plant showing lower stem leaves (but not the basal leaves); also, branch with upper leaves, and flower heads. *a.* Flower head with surrounding bracts and five-lobed tubular flowers. *b.* Achenes with remnants of bristles.

**285**

## RUBBER RABBITBRUSH, big rabbitbrush
SUNFLOWER FAMILY — Compositae

### RUBBER RABBITBRUSH — *Chrysothamnus nauseosus* (Pallas) Britt.

DESCRIPTION — Large woody shrubs, 1½ to 6 feet high with numerous shreddy-barked, flexible trunks from the base. The many young branches and stems arising annually from these are covered by a dense feltlike coat of white to dull white to yellowish wool. These are somewhat sticky due to a resinous gum which gives the plant a characteristic odor. The narrow, alternate leaves, 1 to 3 inches long, less than ⅓ inch broad, are smooth-margined and covered when young by the same feltlike coat of wool.

The small, narrowly coneshaped flower heads, about ¼ to ½ inch high, are in rounded clusters at the ends of the stems. The firm, pointed bracts surrounding each flower head have a resinous-thickened center, and are arranged in 3 to 5 vertical rows. Each flower head contains only 4 to 6 tiny yellow tubular flowers, the tube about ¼ to ½ inch long. Outer ray flowers are lacking in these heads. The slender achenes are angled, and vary from smooth to densely hairy. Each bears a tuft of dull white hairs of unequal length at the top.

There are nine different varieties of rubber rabbitbrush in Arizona. They show great variability in the density of hairs, the shape, and the length of leaves, achenes, and bracts surrounding the flower heads.

DISTRIBUTION — Rubber rabbitbrush is a native weed of deteriorated rangelands. It often grows in dense stands, replacing good forage grasses which have been killed out by overgrazing. Found along roadsides, waste places, or in dry washes, plains, foothills, and bottomlands of mountain valleys. It thrives in alkali and heavy clay, or in sandy, gravelly soil; in sagebrush, pinyon-juniper, and ponderosa pine zones.

Spread throughout Arizona in one or more of its forms, except in the western and southwestern portions; from 2,000 to 8,000 feet elevation; flowering from July to November. Of little forage value to livestock, and unpalatable because of its resinous substances.

POISONOUS PROPERTIES — Although poisonous to livestock, the plant is not eaten in large enough quantities to cause losses.

Fig. 138. Rubber rabbitbrush *(Chrysothamnus nauseosus)*. Leafy branch with flower heads in clusters. *a.* Flower head showing bracts in vertical rows. *b.* Single tubular flower. *c.* Stigmas.

# CANADA THISTLE

SUNFLOWER FAMILY — Compositae

## CANADA THISTLE — *Cirsium arvense* (L.) Scop. var. *mite* Wimm. & Grab.

DESCRIPTION — A prohibited noxious weed in Arizona, Canada thistle is a slender spiny-leaved perennial with rigid stems, 1 to 4 feet high. Reproducing by seed, and by its deep underground vertical and extensive horizontal roots which enable it to spread rapidly over large areas. These creeping roots are greatly branched, and may extend long distances, frequently giving rise to leafy shoots. The many alternate leaves are stalkless, oblong or lanceshaped, and divided into very irregular lobes with many small yellow spines.

The small flaskshaped flower heads are about ½ to ¾ inch in diameter, each containing many small rose-purple tubular flowers. The many, overlapping bracts enclosing each head are spineless. Canada thistle differs from other species of the true thistle in that there are male and female flower heads, and these are on separate plants. Male plants, and thus possible entire colonies, would, therefore, produce no achenes.

The bone or tan colored achenes, about ⅛ inch long, are oblong and smooth, with a tuft of feathery hairs at the top. These hairs fall off after maturity, showing the rounded apex with a little tubercle in the center.

DISTRIBUTION — Canada thistle is an European introduction, and is not from Canada. It now covers the northern half of the United States from the Atlantic to the Pacific, but is only occasional in the southern part. It is one of the most dreaded weeds known to agriculture, as it grows in all cultivated crops, fields, and waste places. Fortunately, it is rare in Arizona. The first established record in Arizona is from Flagstaff in 1920. It is now occasional in Coconino County, and in the Prescott area, Yavapai County.

Fig. 139. Canada thistle *(Cirsium arvense)*. Leafy plant with flower heads.
*a.* Underground roots. *b.* Flower head. *c.* Achene with a tuft of feathery hairs.
*d.* Achene without tuft of hairs, showing blunt apex with a tubercle in the center.

# HORSEWEED

SUNFLOWER FAMILY — Compositae

## HORSEWEED — *Conyza canadensis* (L.) Cronq.
## (*Erigeron canadensis* L.)

DESCRIPTION — An erect annual or biennial, the stems slender to very coarse, usually branching only in the flower part, ¼ to 4 or more feet high; reproducing by seeds. The plants are thinly rough-hairy to practically hairless. The dark green leaves are alternate, 1 to 4 inches long, and ¼ to ½ inch broad, often so crowded as to nearly hide the stem. They are stalkless or the lowermost short stalked, lance- or strapshaped, the margins smooth or more often wavy to few-toothed.

The greenish-white flower heads are small, innumerable in a long, many-branched flowering mass (panicle) at the top of the plant, which is ½ to 2 feet long. Individually they are inconspicuous, about ⅛ inch high and ¼ inch across, with many whitish ray flowers ("petals"), too small to be seen without the aid of a lens. The narrow achenes are tan colored, flattened, 1/25 to 1/16 inch long, with a few scattered hairs on the surface and a thin tuft of about a dozen fine colorless hairs at the blunt end, which are deciduous after maturity.

DISTRIBUTION — Horseweed is introduced from the eastern United States, preferring the rich moist soil of cultivated lands or sandy alluvial soil, but growing in any type of disturbed soil. It is widespread throughout the entire state, but is a pest in the agricultural valleys in cultivated fields, along ditchbanks, row ends, and borders. Also common along roadsides, pastures, yards, gardens, waste places, streams, and hillsides; 100 to 7,500 feet elevation; flowering July to October.

The leaves and flowers of horseweed are reported to contain a terpene which may cause dermatitis and throat irritation in susceptible individuals, and irritation in the nostrils of horses. At one time an oil was distilled from this plant which was used medicinally in the treatment of dysentery and diarrhea.

Fig. 140. Horseweed *(Conyza canadensis)*. Lower portion of plant, unbranched at base, the leaves numerous; upper portion many-branched flowering portion with tiny flower heads. *a.* Enlarged leaf. *b.* Achene with thin tuft of hairs. *c.* Hairy achene, enlarged, without tuft of hairs.

# BROOM SNAKEWEED

SUNFLOWER FAMILY — Compositae

## BROOM SNAKEWEED — *Gutierrezia sarothrae* (Pursh) Britt. & Rusby

DESCRIPTION — Low compact perennial half-shrub, ½ to 1⅓ (or 2) feet high, from a woody base which becomes a branched crown in old plants, reproducing by seeds only. The innumerable stems are very slender and much-branched, woody only at the base. The slender alternate leaves are very numerous on the stems, mostly ⅜ to 1½ inches long, and 1/25 to 1/12 inch broad, with smooth margins.

The yellow flower heads are very small, covered by sticky resin, distinctly top-shaped, broader above and tapering to a narrower base, ⅛ to ¼ inch long, and about ⅛ inch in diameter. They have 3 to 8 deep yellow short marginal (ray) flowers which roll up in age, and 3 to 8 central (disk) flowers. The flower heads are stalkless, or on very short stalks in small dense clusters at the tips of the many small branchlets. Each flower head produces 6 to 16 achenes. These are narrowly oblong, about 1/25 inch long, hairy, and bear at the large end 8 to 12 narrow white scales, 1/25 inch long.

DISTRIBUTION — A native and serious range pest growing well on a wide range of soil sites from gravelly shallow immature soils to deep, sandy, well-developed loams and clayey soils. Common on rangelands from Apache to Mohave counties; at elevations from 2,800 to 8,000 feet; flowering from July to November. It is an aggressive and obnoxious weed, covering millions of acres in northern Arizona. It increases markedly on open grassland and woodland ranges with continued heavy grazing use by livestock and improper range management. Increase in abundance is generally accompanied by loss in grazing capacity.

## THREADLEAF SNAKEWEED — *Gutierrezia microcephala* (DC.) Gray

DESCRIPTION — Very similar in appearance, habitat, and distribution to broom snakeweed. It differs principally in having fewer flowers in each flower head, and thus the flower heads are noticeably narrower. There are only 4 to 5 marginal flowers, and 1 to 3 central flowers in threadleaf snakeweed.

POISONOUS PROPERTIES OF SNAKEWEED — Usually unpalatable plants, but livestock may eat the snakeweeds under conditions of forage shortage. Sheep, cattle, and goats are poisoned, and although death may occur, the major effect of poisoning is abortion. Cattle are more prone to abortion than are sheep and goats.

292

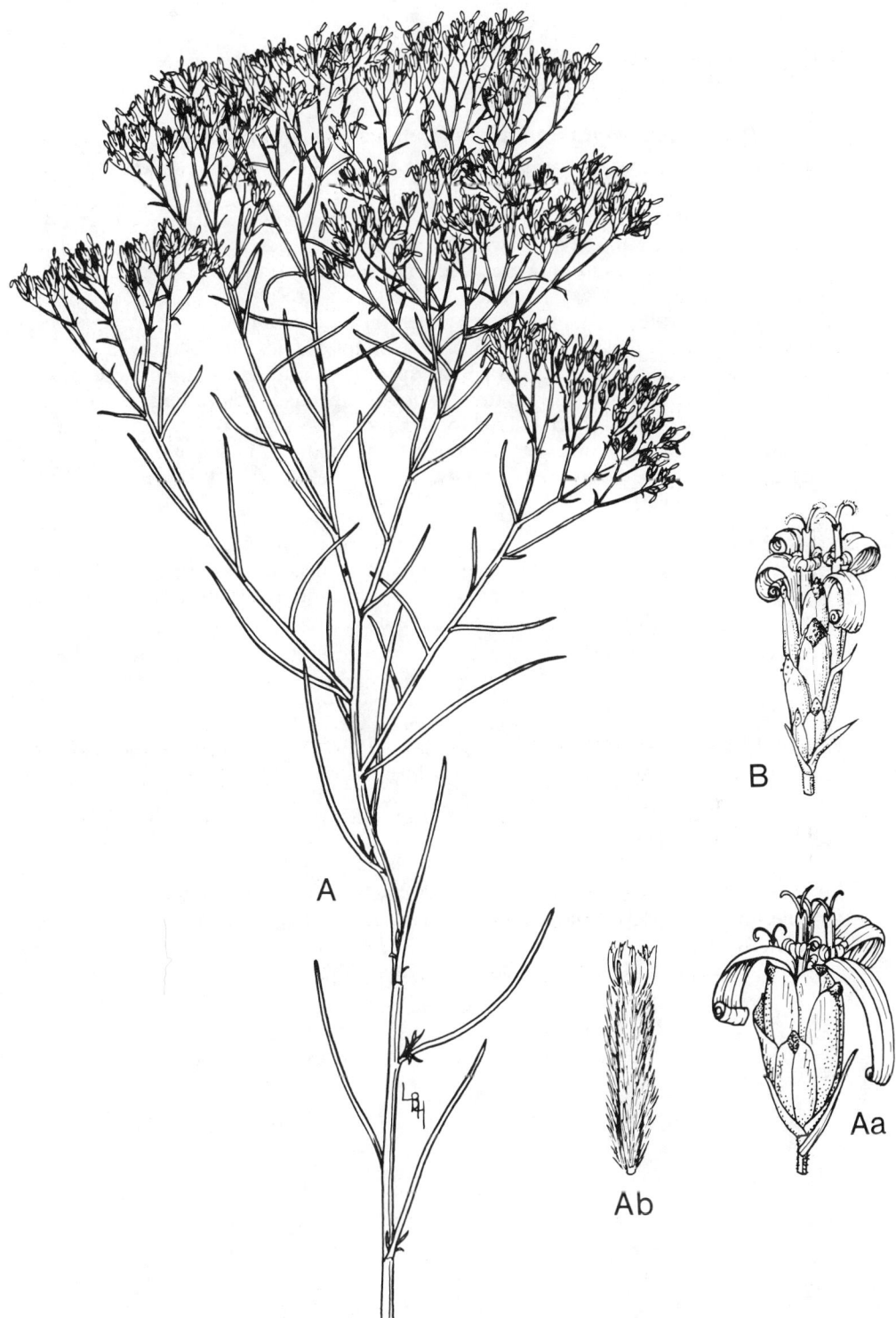

Fig. 141. Broom snakeweed *(Gutierrezia sarothrae)*. *A.* Branch with narrow linear leaves and dense clusters of very small flower heads. *Aa.* Flower head with 3 to 8 central flowers and 3 to 8 marginal flowers. *Ab.* Achene with short pappus. *B.* Threadleaf snakeweed *(G. microcephala).* Flower head narrow, with only 1 to 3 central flowers and 4 to 5 marginal flowers.

293

# BURROWEED

SUNFLOWER FAMILY — Compositae

### BURROWEED — *Haplopappus tenuisectus* (Greene) Blake

DESCRIPTION — Low spreading perennial half-shrub forming rounded bushes, usually the diameter slightly exceeds the height, 1½ to 3⅓ feet across, and ½ to 2½ (or sometimes 3⅓) feet high; reproducing only by seeds. The many-branched stems are woody below. The older and larger ones have gray bark, while the younger ones usually are yellowish-tan. The young stems, leaves, twigs, and flower heads are sticky from resin. The leaves are alternate, ⅔ to 2 inches long, and are divided into 3 to 5 pairs (often obscurely paired) of slender pointed divisions, 1/25 inch broad and 1/25 to ¾ inch long. The yellow flower heads are about ¼ inch long and ⅛ inch broad, but spread when the seeds are mature. They produce only central flowers; the ray flowers, or "petals," are lacking. The flower heads, stalkless or very short-stalked, occur in clusters at the tips of all stems and branches. The achenes are very narrowly topshaped, ⅛ inch long, 4-ribbed or angled, and hairy between the ribs. Arising from the broad end is a tuft of tawny hairs ¼ inch long.

DISTRIBUTION — Burroweed is a native perennial growing on dry, gravelly to sandy soils on roadsides, waste places, alluvial plains, mesas, and slopes. It is a serious range pest on the desert and desert grassland range of eastern Arizona south of the Gila River in Greenlee, Graham, Cochise, Santa Cruz, Pima, Gila, Pinal, and Maricopa counties; 2,000 to 4,500 feet elevation; flowering August to October. Burroweed, worthless as forage, occupies extensive range areas in southern Arizona; this condition is probably the result of too heavy grazing, drought, protection from fire, or a combination of these factors. Reduction in grazing capacity is serious, and always takes place as the plant invades or increases in abundance.

POISONOUS PROPERTIES — Burroweed ordinarily is not eaten by livestock, but animals may be forced to do so in times of forage shortage. Sheep, cattle, and horses are poisoned due to trementol, an alcohol. The most characteristic symptom is trembling, which may shake the whole body. Suckling animals may be poisoned by their mother's milk. Humans may develop milksickness by drinking milk from poisoned cows.

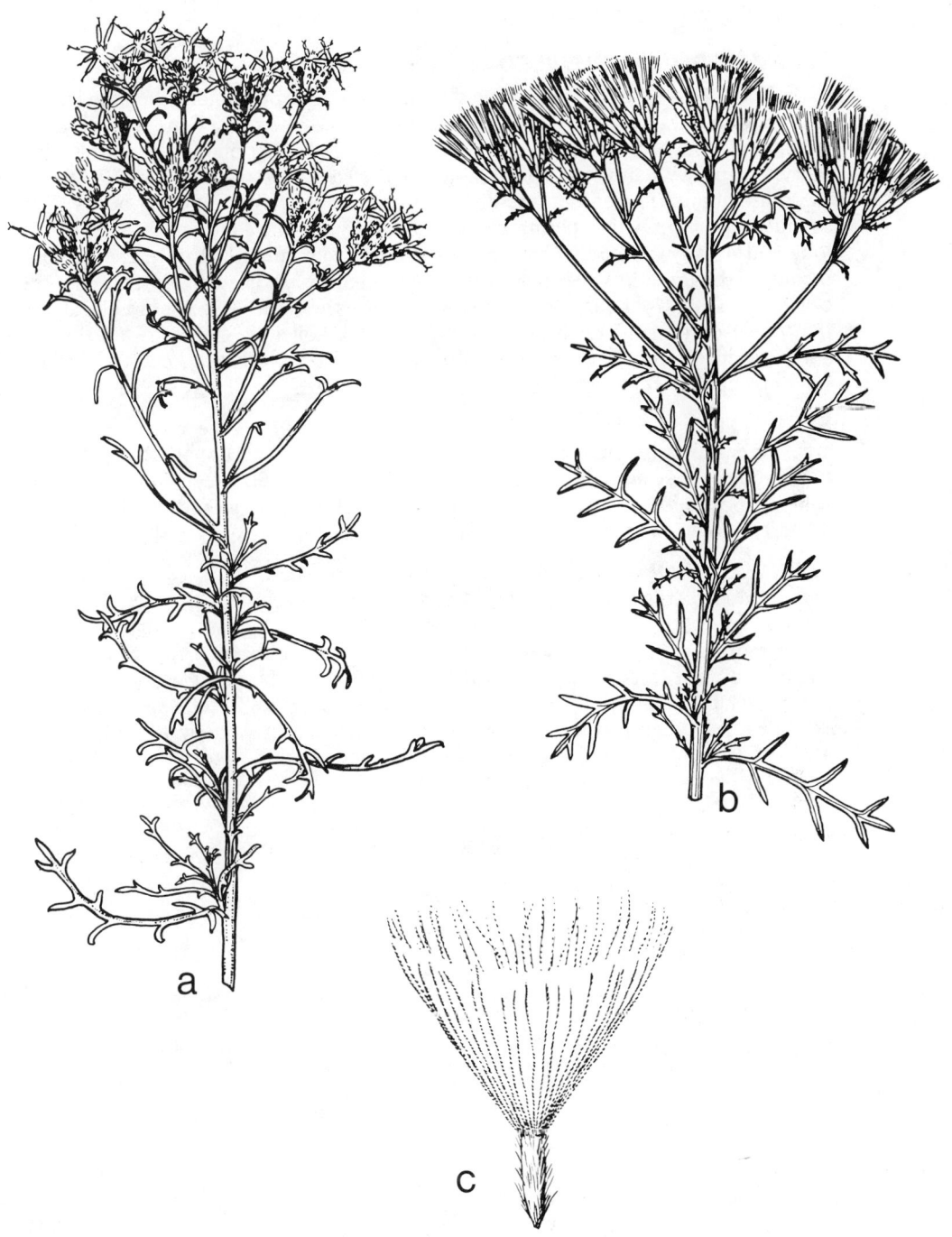

Fig. 142. Burroweed *(Haplopappus tenuisectus)*. *a.* Leafy branch with flower heads.
*b.* Branch with fruits. *c.* Achene showing tuft of hairs which aids in its distribution.

# WESTERN SNEEZEWEED, sneezeweed, orange sneezeweed
SUNFLOWER FAMILY — Compositae

## WESTERN SNEEZEWEED — *Helenium hoopesii* Gray

DESCRIPTION — Western sneezeweed is an herbaceous perennial with large leaves, from a woody taproot, reproducing only by seeds. There are 1 to several stems which are branched above, 1½ to 3 feet high. The young plants are densely soft hairy, but nearly hairless in age. The alternate leaves are thick and stalkless with smooth margins. The ones at the base of the plant are larger, 5 to 15 inches long, rounded at the tip with tapering bases, while those on the stem are 3 to 10 inches long and usually pointed at the tip. The flower heads are 1⅔ to 3 inches across, including the large bright orange-yellow marginal (ray) flowers or "petals." The central (disk) flowers are orange-brown. There are about 3 to 6 flower heads at the top of each stem. Each head produces many achenes, which are topshaped, about ⅛ inch long, densely hairy, ribbed, and bear 5 to 8 thin, colorless membranous scales at the larger end.

DISTRIBUTION — This stout native range weed occurs in the higher mountain areas of eastern Arizona; in Apache, Coconino, Greenlee, Graham, Cochise, and Pima counties; at elevations ranging from 7,000 to 11,000 feet; flowering from June to September. It is found on the summer ranges on moist well-drained mountain meadows, and deep rich soils of coniferous forests.

POISONOUS PROPERTIES — Sneezeweed is poisonous to sheep, cattle, and horses. Most losses occur in sheep. It contains the poisonous glucoside, dugaldin. Although all parts of the plant are poisonous, the fresh plants are more poisonous than the dried ones. The poison is cumulative, and under normal range conditions the animal must feed on the plant about 10 to 20 days before it becomes sick. It causes what is commonly known as "spewing sickness," from the symptoms exhibited. Although this weed is not abundant enough in Arizona to be a serious problem (as in Colorado), some losses may occur in this state.

Fig. 143. Western sneezeweed *(Helenium hoopesii)*. *a.* Basal portion of plant. *b.* Basal
leaf. *c.* Upper part of plant, showing three flower heads with large marginal (ray) flowers
and small central flowers. *d.* Central (disk) flower showing ribbed achene with crown of scales.

297

## SUNFLOWER, annual sunflower, common sunflower
SUNFLOWER FAMILY — Compositae

### SUNFLOWER — *Helianthus annuus* L.

DESCRIPTION — Sunfllower is a tall, robust branched annual, the coarse, rough stems 3 to 7 feet or more high, reproducing only by seed. The rough-hairy leaves are eggshaped or heartshaped, pointed at the tip, the edges usually toothed. They are mostly alternate, 2 to 13 inches or more long, ½ to 6 inches broad, the upper with short stalks, and the middle and lower with stalks to 8 inches long.

The large flower heads are 2 to 5 inches broad, including the bright-yellow ray or marginal flowers ("petals"). The many small central (disk) flowers are tubular, reddish brown, and produce the achenes. The wedgeshaped achenes, about ⅛ to ½ inch long and ⅛ to ⅓ inch broad, are short-hairy only at the summit. They are dark gray with blackish spots and lighter stripes, and produce 2 thin scales at the summit which are deciduous at maturity.

DISTRIBUTION — Sunflower, introduced in Arizona, is native in the Great Plains area. Abundant in moist soils throughout most of the state along roadsides, waste places, abandoned fields, lowlands, barren spots, and ditchbanks where it is often showy and ornamental. Sometimes a serious pest in cultivated crops of sorghum, cotton, citrus orchards, alfalfa, and small grains; 100 to 7,500 feet elevation; flowering March to October or November.

### PRAIRIE SUNFLOWER — *Helianthus petiolaris* Nutt.

DESCRIPTION — Very similar in appearance to the sunflower, but generally more slender and shorter, mostly ½ to 3 feet high. The leaves are smaller, narrower, more lanceshaped, and the edges are not usually toothed. The flower heads are smaller, 1 to 2 inches broad with stiff whitish hairs visible in the dark centers. The same general distribution as sunflower, but not as common.

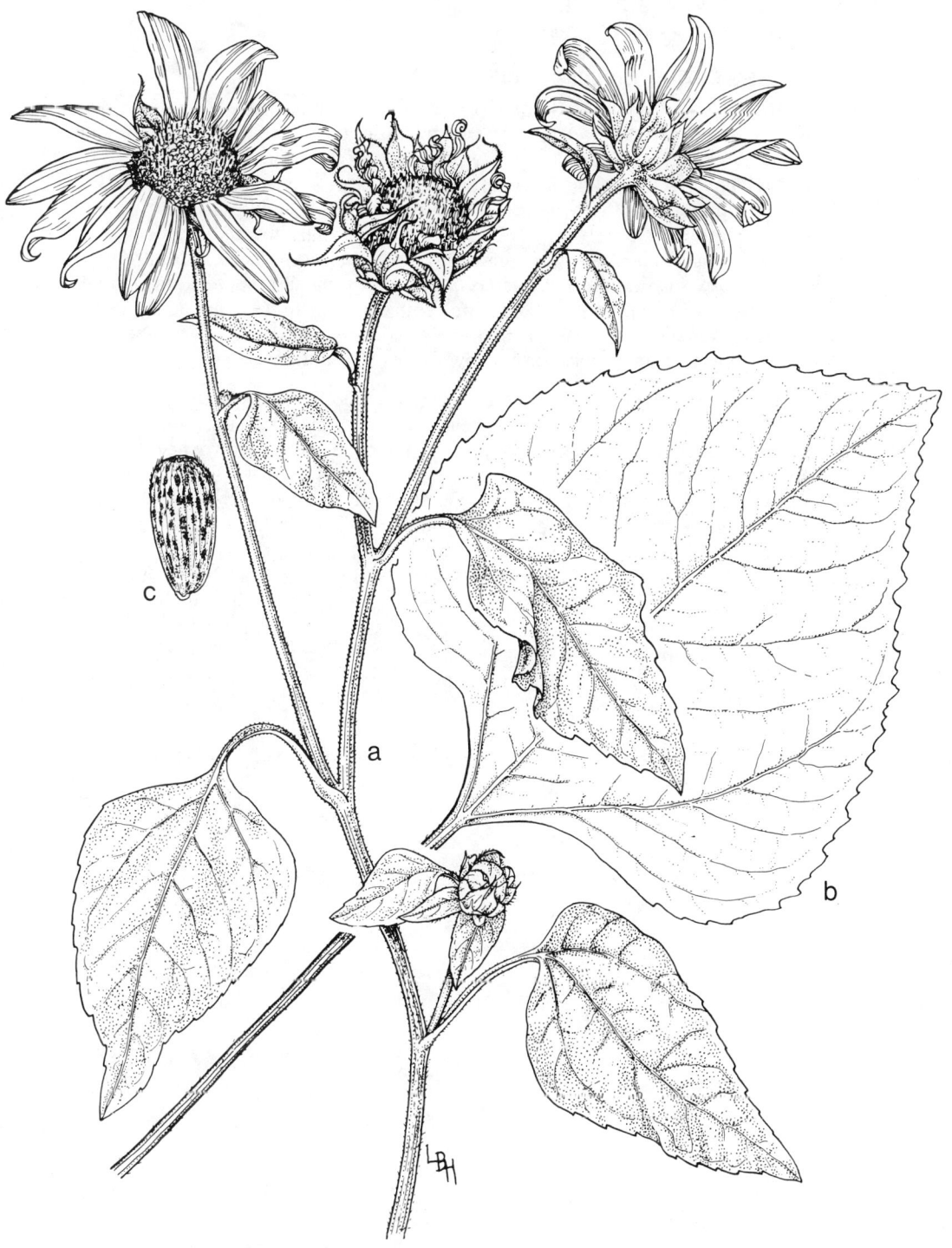

Fig. 144. Sunflower *(Helianthus annuus). a.* Upper part of plant with showy
flower heads and leaves. *b.* Large basal leaf. *c.* Achene with dark spots
and slightly hairy summit; the two scales have dropped off.

# BLUEWEED, yerba parda
### SUNFLOWER FAMILY — Compositae

## BLUEWEED — *Helianthus ciliaris* DC.

DESCRIPTION — A prohibited noxious weed in Arizona, blueweed is a low bluish or grayish green perennial, 1 to 2 feet high; reproducing by seed, but principally by the widespreading underground system of upright roots and rhizomes. The leaves, which give the plant its characteristic color, are stemless, and may be alternate or opposite. They are narrow to broadly lanceshaped, 1 to 4 inches long, and vary greatly in width, from 1/12 to ¾ inch. The margins are often very wavy, and bear short stiff hairs. Similar hairs may also occur along the veins on the lower surface; otherwise, the plant is hairless. The flower heads, ½ to 1 inch broad, have yellow ray or marginal flowers. The central (disk) flowers are dark purplish brown. The achenes are grayish brown, 4-angled, somewhat wedgeshaped, and about ⅛ inch long, often with dark brown spots at maturity.

DISTRIBUTION — Blueweed is one of the few noxious weeds in the state that is native. It grows in dense colonies in heavy saline or alkaline soil, in sandy loam ditchbanks, in low drainage areas, dry lakes, and roadsides from whence it spreads to cultivated lands. It is found nearly throughout the state from 100 to 7,000 feet elevation, and flowers from May to October. It is especially troublesome in southern Apache and Navajo counties in the White Mountains area, and in the farm areas of Graham, Cochise, Pinal, and Pima counties.

Blueweed is a potential pest in any cultivated field it invades. Cultivation may aid its spread. The rhizomes are cut into pieces and distributed throughout the soil. These pieces are capable of starting new plants. Its growth is so persistent that many crops cannot compete favorably, and eventually may be crowded out. A circular patch of blueweed, 500 feet in diameter, was observed in a bean field near Vernon (Apache County).

Fig. 145. Blueweed *(Helianthus ciliaris)*.
Top of plant showing leaves and flower
heads.  *a.* Achene mottled dark brown.

# CAMPHORWEED

SUNFLOWER FAMILY — Compositae

## CAMPHORWEED — *Heterotheca subaxillaris* (Lam.) Britt. & Rusby

DESCRIPTION — Tall coarse hairy annual or sometimes biennial, with a strong odor, from a taproot; reproducing by seeds only. There is a single principal stem, mostly branching only at the top and spreading, resembling a telegraph pole, 2 to 6 feet high. It is more or less covered by long, spreading hairs with the upper branchlets and flower stalks bearing gland-tipped hairs filled with a sticky secretion.

The leaves are alternate with toothed margins. The lower and basal leaves are large, oval, or oblong; they are on a slender stalk, and usually there is a pair of leaflike lobes on either side of the stalk base. The upper leaves are much smaller and stalkless, their bases heartshaped and clasping the stem.

The flower heads are yellow, about ¼ to ⅜ inch high, and ½ to ¾ inch in diameter, including the 20 or more deep yellow petallike ray or marginal flowers. The 35 or more central (disk) flowers are also yellow. The achenes are of 2 kinds, both of which are maroon colored and 1/12 to ⅛ inch long. The ray achenes are 3-angled, entirely hairless and without the apical tuft of hairs. The achenes of the central flowers are covered by whitish hairs, not 3-angled, and have a tuft of brownish hairs at the larger end.

DISTRIBUTION — Camphorweed, a native of tropical America, grows in moist or dry sandy soil along ditchbanks in cultivated fields, or edges, roadsides, and low places where water collects, also in sandy washes. Although it grows in dry soil, it attains its best growth in moist soil, as in and around the irrigated lands. Common throughout central and southern Arizona, 100 to 5,500 feet elevation; flowering March to November, but largely in late summer.

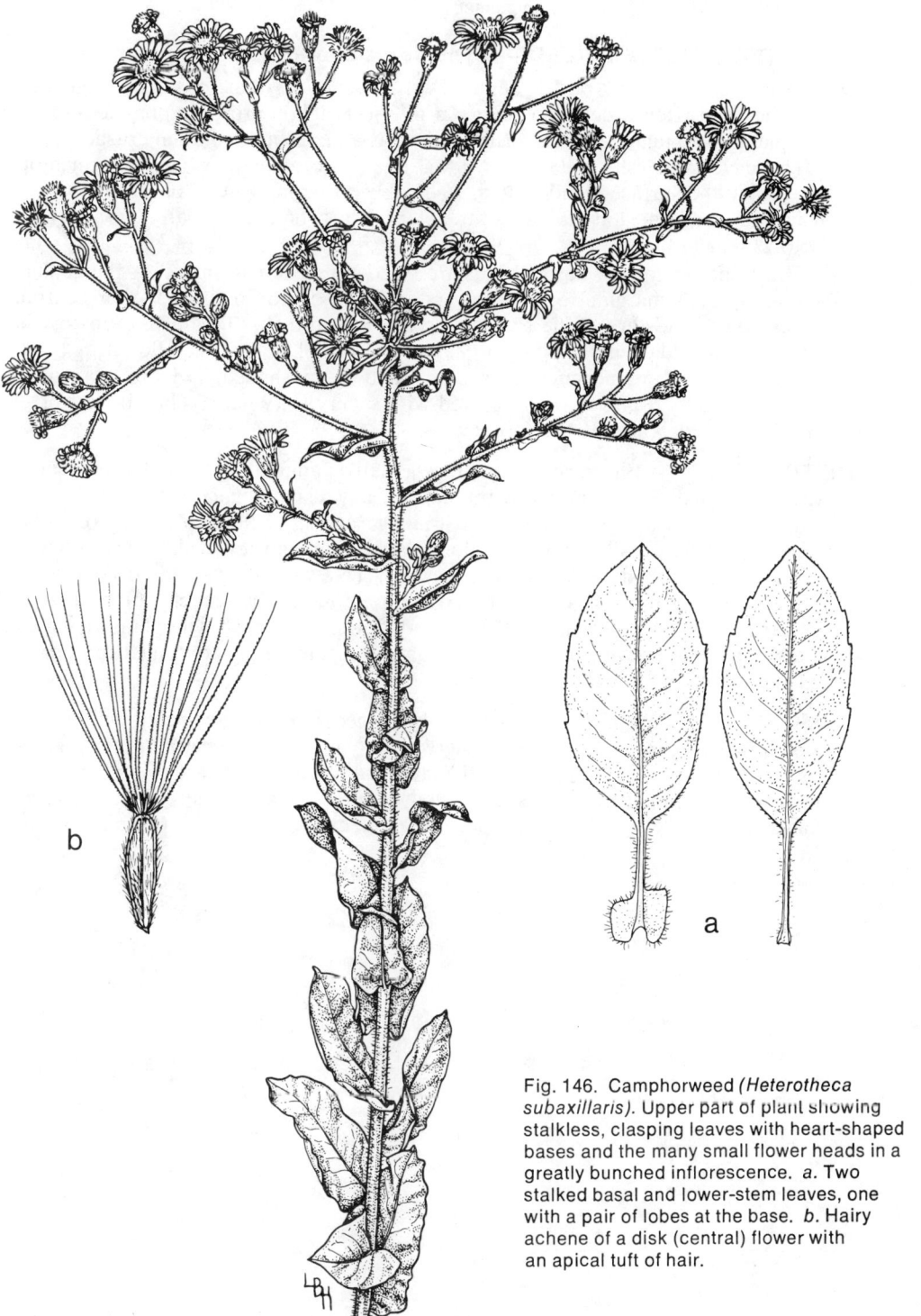

b

a

Fig. 146. Camphorweed *(Heterotheca subaxillaris)*. Upper part of plant showing stalkless, clasping leaves with heart-shaped bases and the many small flower heads in a greatly bunched inflorescence. *a.* Two stalked basal and lower-stem leaves, one with a pair of lobes at the base. *b.* Hairy achene of a disk (central) flower with an apical tuft of hair.

303

# BITTER RUBBERWEED, Bitterweed
## SUNFLOWER FAMILY — Compositae

### BITTER RUBBERWEED — *Hymenoxys odorata* DC.

DESCRIPTION — Low, bushy annual with several or numerous stems, greatly branched and often widely spreading, ⅓ to 2 feet high, reproducing by seeds only. The plant, like pingue, is bitter tasting, and has a pungent odor when crushed.

The leaves are fleshy and divided into 3 to 13 threadlike divisions, 1/16 inch or less broad, the surface gland dotted. A quickly-withering rosette of leaves, 1 to 4 inches long, is formed at the base of the plant. Those on the stem are alternate and ¾ to 2¼ inches long.

The numerous yellow flower heads are ⅓ to 1 inch across, including the golden yellow, 8 to 13 marginal or ray flowers, and the 50 to 75 tiny disk, or central flowers. The heads are borne singly on stalks 1 to 6 inches long at the tips of all the branches, and a single plant may have hundreds of heads. Each flowering head produces 50 or more achenes. These are narrowly topshaped, 1/16 to 1/12 inch long, indistinctly 4-angled, and covered with silvery silky hairs. They bear 5 or 6 whitish pointed scales at the top.

DISTRIBUTION — Bitter rubberweed is a native annual range weed of southwestern United States, growing in moist heavy clay, alkali, adobe, sandy, or alluvial soils. It is most common in drainage areas, as flood plains, lake beds, roadsides, and especially abundant along the bottom lands of the lower Gila River (Yuma County), and Little Colorado River (Navajo and Coconino counties). Found in southern and north central Arizona; 350 to 6,000 feet elevation; flowering January to June, but in moist spots may flower until frost. The plant increases in abundance whenever the grass cover has been thinned by continued too heavy grazing use or by drought.

POISONOUS PROPERTIES — Bitter rubberweed is ordinarily eaten only when good forage is scant. Since it becomes green early in the spring before other vegetation, animals hungry for green feed may be attracted to it. Mature plants and those growing in dry situations are usually more poisonous than young vigorous plants. Although reported to be poisonous to cattle, it is most troublesome to sheep.

304

Fig. 147.  Bitter rubberweed *(Hymenoxys odorata)*.  Young plant showing flower
heads and finely lobed leaves; the marginal flowers are persistent and droop
at maturity.  *a.* Achene with silky hairs and five scales at the top.

## PINGUE, Colorado rubberweed
### SUNFLOWER FAMILY — Compositae

### PINGUE — *Hymenoxys richardsoni* (Hook.) Cockl. var. *floribunda* (Gray) Parker

DESCRIPTION — Low, tufted, upright perennial, from a branched woody root crown, with long cottony hairs in the old leaf bases, reproducing by seeds only. The plants are bitter tasting, with a strong pungent odor. The several stems are greatly branched above, and 6 to 18 inches high.

The fleshy leaves, with the surface gland dotted, are rather rigid. They form basal tufts around the stems, are 1½ to 6 inches long, with a dense mass of cottony hairs between the axils. The leaves on the stem are alternate, ¾ to 2 inches long. All are very narrow, and may be undivided or mostly divided into 3 to 7 long narrow lobes, about 1/25 to 1/12 inch broad.

The yellow sunflowerlike heads are ½ to 1¼ inch across, including the 7 to 10 bright yellow marginal or ray flowers, and the tiny, tubular central flowers. The ray flowers are persistent, and become whitish and pendent in age. The heads occur in large numbers at the tips of all the branches, giving the plant a flattopped appearance. Each head produces many achenes similar to those of bitter rubberweed, but about ⅛ inch long.

DISTRIBUTION — A native perennial weed, thriving on all soils from dry to moist, poor to fertile, and sandy to heavy clays. Pingue reaches its greatest development in open grasslands, occupying all slopes and exposures with nearly equal vigor and density. It grows well, although stands may be sparce under oak, juniper, and ponderosa pine. A serious and abundant weed on northeastern Arizona ranges, from Apache to Coconino and Yavapai counties; 5,000 to 9,000 feet elevation; flowering from late June to September.

Pingue increases in abundance whenever the grass cover is thinned by continued close grazing. The plant contains a high percentage of latex, which was extracted during both World War I and II, and manufactured into rubber. High costs of harvesting and latex extraction, and slow growth of the plant prevent commercial development.

POISONOUS PROPERTIES — Pingue is poisonous to sheep. The plant is not eaten by cattle. Death losses in sheep may occur at any time of the year, but the greatest danger is in the spring or late fall when palatable forage is apt to be scarce, and sheep are forced to eat pingue.

Fig. 148.  Pingue *(Hymenoxys richardsoni* var. *floribunda).* Leafy
branch from the woody crown, showing flower heads and narrowly-divided
leaves.  *a.* Central, or disk flower with its achene, crowned by scales.

# POVERTYWEED

SUNFLOWER FAMILY — Compositae

### POVERTYWEED — *Iva axillaris* Pursh

DESCRIPTION — Povertyweed is a rank smelling, very leafy, bushy perennial, usually growing in colonies, and reproducing by creeping roots, erect roots, and seed. The slender stems, branching from the base, are ½ to 2 feet high. The small, thick, stalkless leaves, opposite below and alternate above, are oblong, ½ to 1½ inches long, and grow to the tip of the branches.

Small drooping flower heads enclosed by a green cup hang downward from short stalks. They are composed of tiny flowers; the outer 5 are fertile and female; the inner 10 to 25 are sterile and male. The heads occur singly in the leaf axils on the upper half of the stems, and each produces 5 fruits. The brown, topshaped, granular achenes, 1/12 to ⅛ inch long, are beaked at first.

DISTRIBUTION — Povertyweed is a native plant, preferring alkaline soil. It has become a pest in grain fields, other crops, flats, and waste places in many areas in northern Arizona (e.g. Woodruff, Show Low, Tuba, Pumkin Center, Fredonia), in Apache, Navajo, and Coconino counties; 4,000 to 7,500 feet elevation; flowering May to September. Once this weed becomes established, it may spread at an alarming rate.

### MARSHELDER — *Iva xanthifolia* Nutt.

DESCRIPTION — A tall robust annual, 2 to 6 feet or more high, reproducing by seeds. The long stalked, large leaves, mostly opposite, are 3 to 12 inches long, and similar to those of the cocklebur. The flowers are like those of povertyweed, but here the flower heads are stalkless, and are crowded on long, branching spikes at the top of the stems and the upper leaf bases. The flowering part of the plant may be greatly branched and spreading, and 1 to 2 feet or more long. The achenes are blacker than those of povertyweed, and not granular, but otherwise similar.

Marshelder grows in moist soil along streams, roadsides, and waste places in northern Arizona, and is also an infrequent garden weed in the Phoenix area in Apache, Navajo, Yavapai, and Maricopa counties; 1,000 to 6,500 feet elevation; flowering July to October. The pollen may cause serious hay fever, and the leaves produce a skin rash in some people. Fortunately, it isn't very widespread in Arizona.

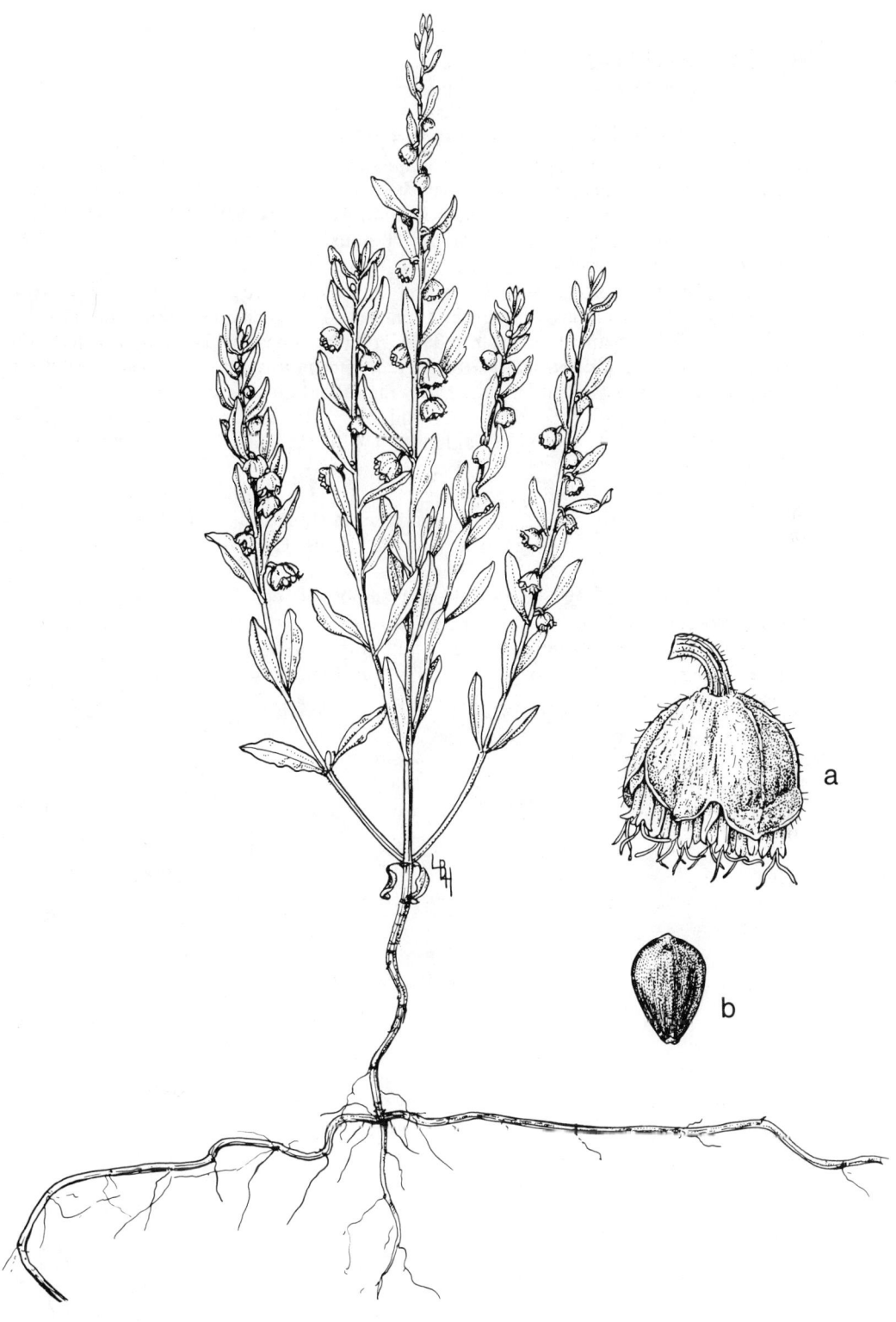

Fig. 149. Povertyweed *(Iva axillaris)*. Leafy plant showing creeping roots and flower heads on short pendulous stalks, arising in the leaf axils. *a.* Enlarged flower head. *b.* Achene.

309

# PRICKLY LETTUCE

SUNFLOWER FAMILY — Compositae

## PRICKLY LETTUCE — *Lactuca serriola* L.

DESCRIPTION — Stiff erect annual, winter annual, or biennial with milky juice, from a large taproot, reproducing only by seeds. There is 1 stiff, principal stem, often hollow, branching only in the flowering part, or sometimes with a few short branches from the base, smooth above, but usually with sharp prickles on the lower part, 2 to 6 feet high.

The leaves are alternate, bluish green, stalkless, tightly clasping the stem with 2 angled or earlike lobes. The lower leaves are 2 to 10 inches long, and are of 2 forms. The margins may be cut into deep irregular lobes, or they may be unlobed, as in the common var. *integrata* Gren & Godr. Both types often have prickles along the margins, the large white midvein, and the other veins on the lower leaf surface. The upper stem leaves are all smaller, unlobed, and similar to the second type of lower leaves. Seeding leaves are unlobed, and form a basal rosette at the ground level.

There are many yellowish flower heads, ⅛ to ⅓ inch broad, composed entirely of petallike ray flowers. The flower heads are on short stalks, and are borne at the top of the plant on many branched flowering stems. Each flower head produces 6 to 30 achenes.

The achenes are eggshaped, flattened, about ⅛ inch long or less, light gray, with 5 to 7 parallel ridges on each side, short-bristly near the summit, and a very slender beak arising from the broad end, which bears a tuft of fine white hairs. The hairs drop off quickly.

DISTRIBUTION — Prickly lettuce is an introduced European weed. It grows in any type of disturbed soil. It is a serious pest in irrigated crops, especially in those which are not weeded regularly, as grain and alfalfa fields, orchards, and vineyards. Also common along roadsides, yards, and gardens. Widespread throughout the state; from 100 to about 8,000 feet elevation; flowering May to October.

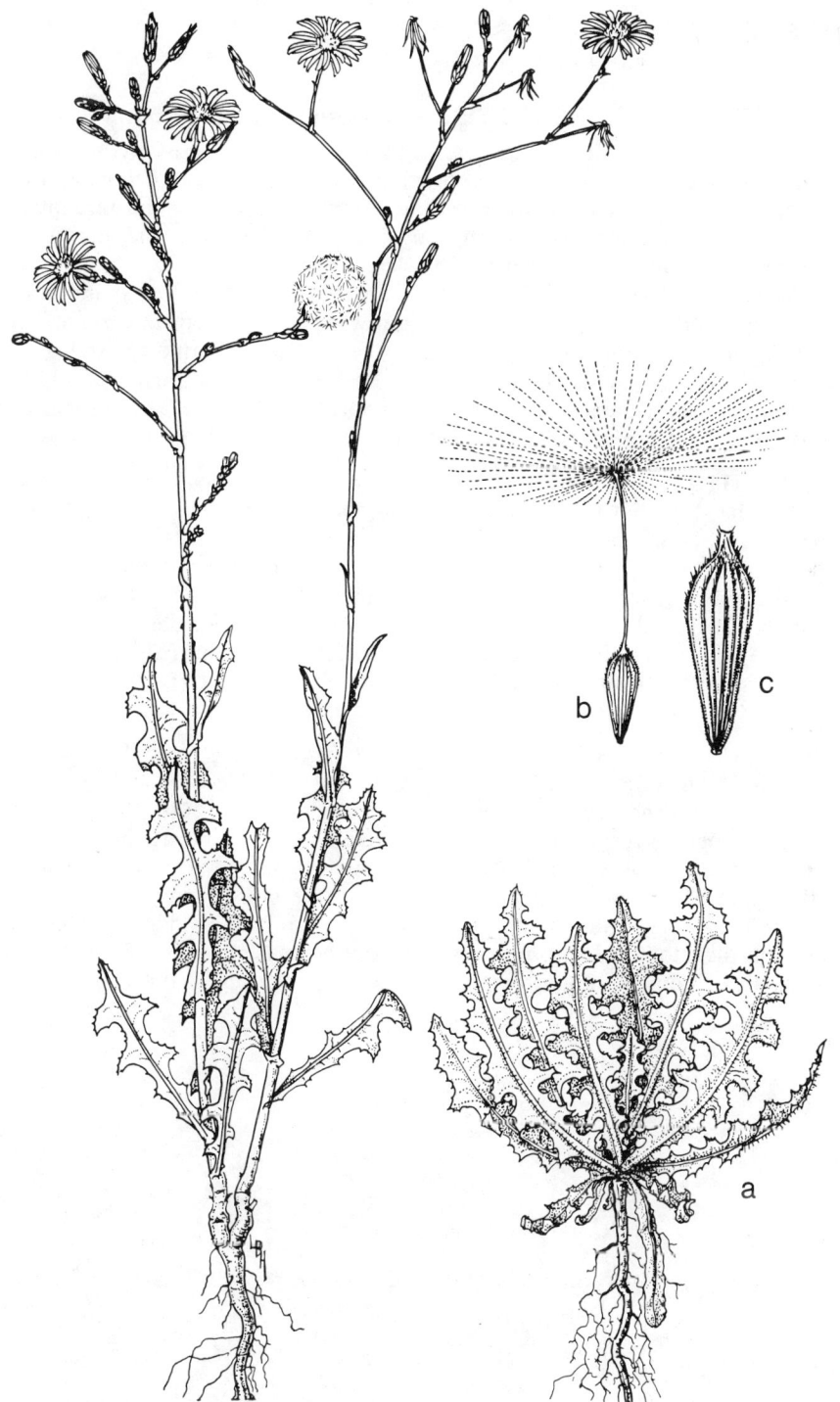

Fig. 150. Prickly lettuce *(Lactuca serriola)*. Plant with flower heads and lobed lower leaves. *a.* Basal rosette leaves of seedling plant.
*b.* Achene with long slender beak bearing a parachutelike tuft of hairs.
*c.* Enlarged achene showing lengthwise ridges and short bristles near summit.

# THREADLEAF GROUNDSEL

SUNFLOWER FAMILY — Compositae

## THREADLEAF GROUNDSEL — *Senecio longilobus* Benth.

DESCRIPTION — A half-shrub, forming scraggly bushes, mostly 2 to 6 feet high; reproducing by seeds. The several branched stems are woody below, and become herbaceous at the tops. The leaves, divided into long narrow and threadlike divisions, are 1½ to 2½ inches long, arranged alternately on the stems, and are permanently white-woolly hairy. The yellow erect flower heads are bellshaped, about ½ inch high and 1½ inch across. There may be several to many in a loose, somewhat flattopped cluster. The outer row of strapshaped flowers, or petallike ray flowers, are about ½ inch or more in length, seed-producing, and 8 to 18 in number. The small inner or disk flowers are yellow, numerous, tubular, and are also fertile. The achenes are narrowly cylindrical, about 1/12 inch long, covered with fine, close-pressed whitish hairs, with a tuft of long white hairs at the tip.

DISTRIBUTION — A native shrubby range weed found mostly in the southeastern half of the state, but is distributed throughout the state from Apache to Mohave, south to Cochise, Santa Cruz, and Pima counties; from 2,500 up to 7,500 feet elevation. It occurs on dry, rocky, or gravelly sites on plains, mesas, and along dry flood plains and watercourses. It seldom grows abundantly, but is common and widely distributed. Associated mostly with creosote bush, desert grassland, and pinyon-juniper woodland vegetation, but extending up into the ponderosa pine belt. Also common along roadsides and waste places; flowering in southern Arizona throughout most of the year, but most abundantly from April to November.

POISONOUS PROPERTIES — During most of the year, threadleaf groundsel is unpalatable to domestic livestock. However, it is evergreen, and begins growth earlier in the year than most other range vegetation. At this time, animals are hungry for green forage, and may be attracted to it. Ordinarily, however, they do not eat it, except in dire necessity. The leaves of the new growth are most toxic. Cattle and horses are equally susceptible and the poison may have a cumulative effect. Poisoning is due to the presence of several alkaloids, and death occurs through their toxic effect on the liver. Under normal range conditions, sheep and goats are seldom poisoned by it to the extent of causing death.

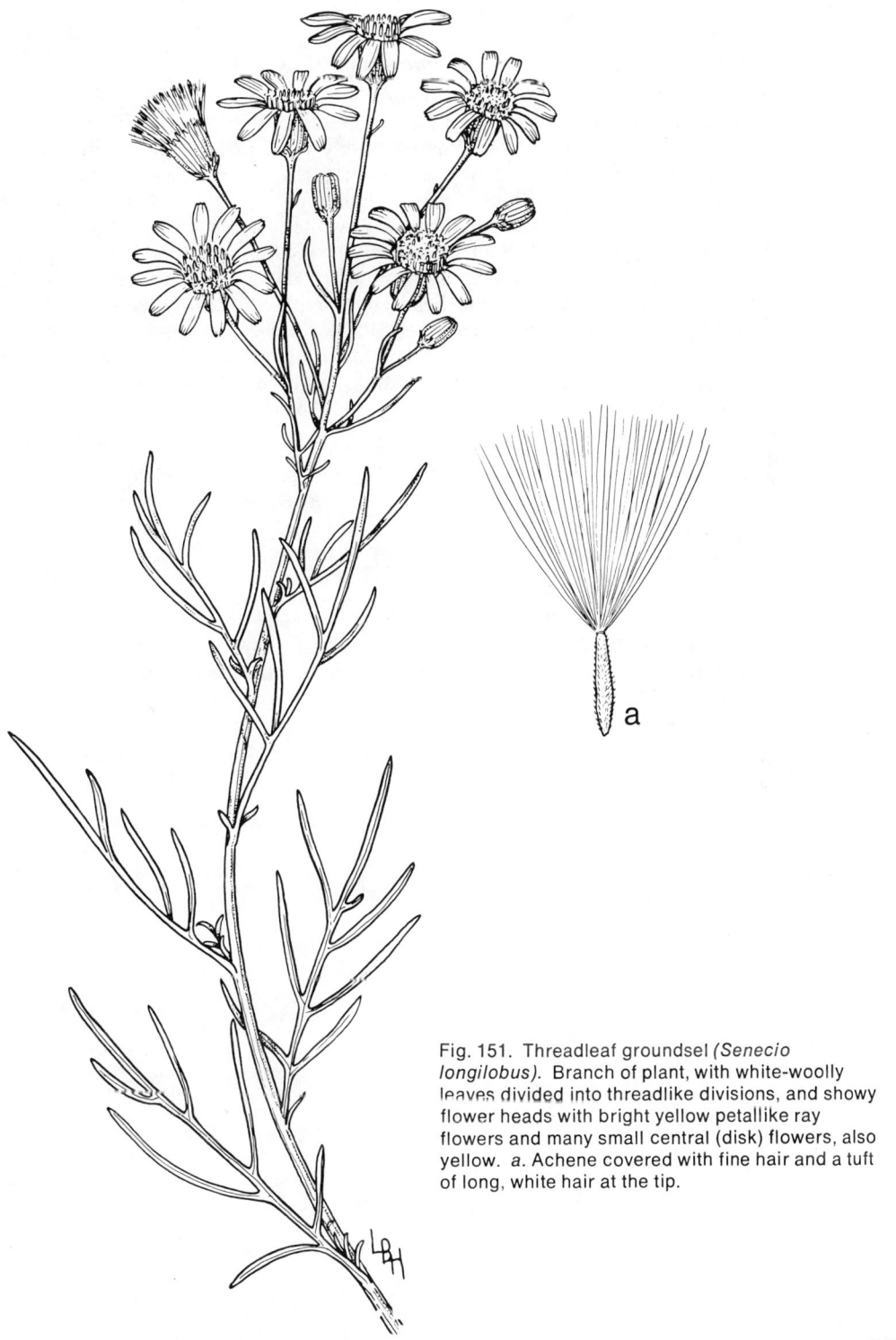

Fig. 151. Threadleaf groundsel *(Senecio longilobus)*. Branch of plant, with white-woolly leaves divided into threadlike divisions, and showy flower heads with bright yellow petallike ray flowers and many small central (disk) flowers, also yellow. *a.* Achene covered with fine hair and a tuft of long, white hair at the tip.

# MILK THISTLE

SUNFLOWER FAMILY — Compositae

## MILK THISTLE — *Silybum marianum* (L.) Gaertn.

DESCRIPTION — Stout spiny annuals or biennials, 2 to 5 feet high, from a thick taproot, reproducing by seed. The leaves are large, the lower deeply-lobed, 1 to 2 feet long and stalked; the upper progressively shorter, fewer-lobed and stalkless, clasping the stem at the base with a pair of earlike lobes. The leaf margins are edged with many yellow spines, ⅛ to ½ inch long. The main veins on the upper leaf surfaces are more or less outlines with white, often there are also scattered, white-blotched patches. The thistlelike purple flower heads are often nodding, globeshaped, 1 to 2 inches broad, and solitary at the end of long stalks. There are no outer petallike ray flowers. The heads are composed of many red-purple central tubular flowers with 5 long narrow lobes. The leathery bracts surrounding each flower head are in several overlapping series, spine-margined, and tipped by stiff spines ½ to 1½ inches long. The achenes are smooth, shiny, mottled with buff and dark brown, and about ¼ inch long. They bear a tuft of white scaly bristles at the tip, which are deciduous in a ring.

DISTRIBUTION — Milk thistle is a native of the Mediterranean region. It is occasional in most areas in southern Arizona, but common in the Salt River Valley. It grows along roadsides, irrigation ditches, pastures, and waste places in Maricopa, Gila, Pinal, Pima, and Cochise counties; mostly from 1,000 to 3,500 feet elevation; flowering March to August, but most abundantly from April to June.

POISONOUS PROPERTIES — Although milk thistle may contain nitrates at a potentially toxic level when growing on fertile soil, it is not common enough in Arizona to cause poisoning. Heavy losses of cattle and sheep from this weed have occurred in Australia, and some in California.

Fig. 152. Milk thistle *(Silybum marianum)*. Plant with spiny, lobed leaves, the veins white-marbled, also showing the globose flower head with the spiny-tipped, overlapping bracts enclosing the head. *a*. Achene with a tuft of scaly bristles.

315

# SPINY SOWTHISTLE

SUNFLOWER FAMILY — Compositae

## SPINY SOWTHISTLE — *Sonchus asper* (L.) Hill

DESCRIPTION — A tall, erect fleshy annual or sometimes biennial, with milky juice from a stout taproot; reproducing only by seeds. The large hollow stem is often reddish, unbranched, or few branched, 1 to 4 or more feet high. The upper stems and flower stalks are sparingly to densely covered with stalked glands in the more common growth form. However, glands may be entirely lacking in some plants.

The alternate leaves are numerous, and usually deeply lobed, with the margins soft prickly-toothed. The lower leaves, up to 12 inches long, are definitely stalked. These and the principal stem leaves are divided into 5 to 11 lobes along each side, with the tip-lobe not longer, nor broadly triangular. Only the lowest leaves are stalked. The middle and upper leaves are stalkless, and clasp the stem with a pair of large rounded earlike lobes. The leaves become progressively smaller and fewer-lobed above. The margins of the upper leaves are irregularly cut, jagged, and long toothed, but usually not divided into definite lobes. However, in some growth forms, even the lower and middle leaves are scarely lobed.

The flower heads are numerous, small, ½ to 1 inch across, and composed wholly of pale yellow petallike ray flowers, and no disk flowers. The bracts surrounding the flower head and the flower stalks may bear stalked glands (gland-tipped hairs). The achenes are reddish-brown, flattened, and margined with a narrow wing; the edges are thinner than the body. They are 2 to 2½ times as long as broad, with the broadest part through the middle, and about 1/12 to ⅛ inch long. There are 3 distinct central ribs (rarely 4 or 5) on each face, with no cross ridging between them, but they may be cross wrinkled near the edges and on the winglike margins. The tuft of fine hairs drops away quickly.

DISTRIBUTION — Spiny sowthistle is a naturalized European weed, widespread throughout the state, but more abundant and serious in the southern part. A pest in grainfields, alfalfa, winter vegetable crops, orchards, lawns, ditchbanks, and waste places. It flowers nearly throughout the year in some places, but principally from November to May; 150 to 8,000 feet elevation.

316

Fig. 153. Spiny sowthistle *(Sonchus asper).* Plant showing stalked basal
leaves, stalkless principal leaves, small flower heads, heads in bud, and head in
fruit. *a.* Base of principal leaf showing pair of rounded earlike lobes.
*b.* Flower head composed wholly of petallike ray flowers. *c.* Achene, enlarged,
with three distinct central ribs and no cross wrinkles except near margin.

# ANNUAL SOWTHISTLE

SUNFLOWER FAMILY — Compositae

## ANNUAL SOWTHISTLE — *Sonchus oleraceus* L.

DESCRIPTION — Tall fleshy annual, very similar to spiny sowthistle, but the achenes and leaves are different. The leaves are very thin, and all (except some uppermost) are deeply-lobed. In the most common form, the principal leaves are deeply cut into 1 to 3 lobes along each side, with the tip-lobe much larger and broadly triangular; or sometimes without side lobes, but just one large lobe at the tip, and a long narrow winged stalk. Unfortunately, the other common leaf form is similar to the many (5 to 7) lobed leaf of spiny sowthistle, with the tip-lobe no larger than the side lobes. However, there is one consistent leaf characteristic — the pair of projections at the base of the stalkless, clasping leaves are sharp-pointed or taper to a point, and are not rounded and earlike as in spiny sowthistle.

The flower heads are slightly larger in annual sowthistle, about ¾ to 1¼ inches broad. No gland-tipped hairs are found in any parts of the plant. The achenes are about the same length, 1/12 to ⅛ inch, but are 3½ to 4 times as long as broad, with the broadest part near the top. There are 5 to 7 indistinct ribs on each face, with roughened, cross-wrinkling in the furrows between them. The edges are not thin nor wing-margined, but as thick as the body. The difference between the achenes are best seen with a hand lens.

DISTRIBUTION — Annual sowthistle is native of North Africa, Europe, and western Asia. Found throughout most of Arizona, but not as abundant as spiny sowthistle, and not a serious pest in winter croplands. Growing in vacant lots, roadsides, and uncultivated grounds as well as cultivated fields, gardens, bottomlands, and ditchbanks. The young leaves of both this plant and spiny sowthistle were gathered in the spring and cooked as greens, but probably an uncommon practice in modern Arizona. Found from 150 to 7,000 feet elevation; flowering February to November.

318

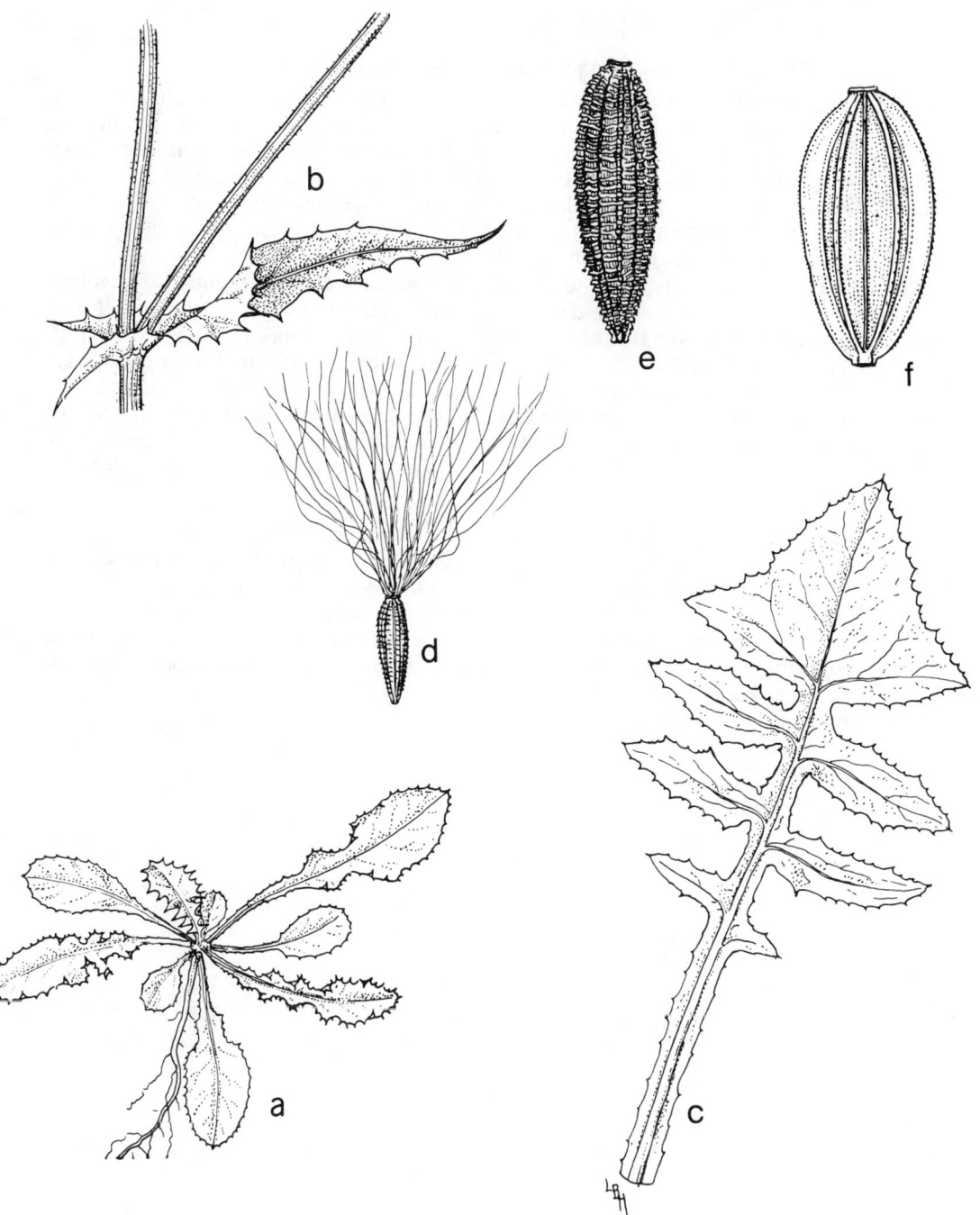

Fig. 154. Annual sowthistle *(Sonchus oleraceus). a.* Principal leaves of young plant showing larger, triangular-shaped, tip-lobe and just 1 to 3 pairs of lobes on each side of leaf. *b.* Attachment of clasping leaf to stem with a pair of sharp-pointed projections. *c.* Stalked basal leaf. *d.* Achene with tuft of fine hairs. *e.* Achene with 5 to 7 indistinct ribs and strong cross wrinkling in furrows between ribs. *f.* Achene of *S. asper* for comparison, both enlarged.

319

# DANDELION

SUNFLOWER FAMILY — Compositae

### DANDELION — *Taraxacum officinale* Weber

DESCRIPTION — A low tufted perennial from a long thick branched taproot, reproducing by seeds and by new shoots from the root crowns. There are no true stems, and the leaves are clustered at the base of the plant. They form a rosette on the ground, or are ascending to upright. The leaves vary greatly in size and lobing. They are from 2 to 12 inches long, and usually divided into few or several indistinct pairs of lobes, which are pointed or blunt at the tips. Often the lobe at the tip of the leaf is much larger, and triangular in shape.

The flower heads are 1 to 2 inches across, and composed entirely of golden yellow petallike flowers or rays. The flower heads are solitary at the end of long naked hollow flower stalks, which are 3 inches to 2 or more feet long. The strap-shaped ray flowers are 5-notched at the tip. There are 100 to 300 in each head. The achenes are greenish or light brown, about ⅛ inch long, 5-to 8-ribbed on each side, and minutely toothed with tiny curved spines along the upper margins. The achene ends in a long slender beak 2 or 4 times as long as the body of the achene, and is attached to a tuft of fine silky whitish parachutelike hairs which are persistent.

DISTRIBUTION — Dandelion is a European introduction. It grows in moist places, and is a much-hated pest in lawns throughout the state. It also grows in the cultivated fields and croplands, along roadsides, in yards, gardens, pastures, and on overgrazed or eroded areas in open mountain meadows of the high mountain ranges, or in moist soil along streams in lower ranges; 100 to 9,000 feet elevation; flowering in some places almost the year around. It is good forage on the ranges, and is especially relished by sheep.

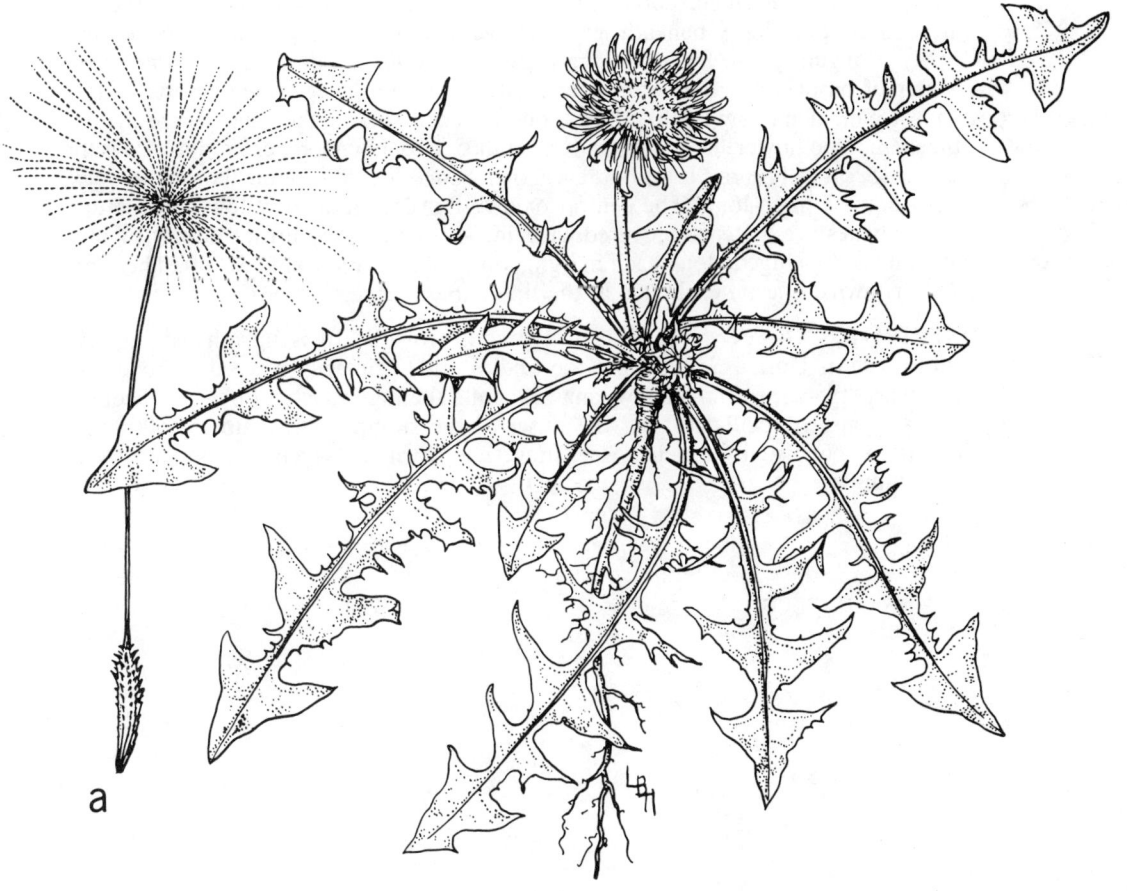

a

Fig. 155. Dandelion *(Taraxacum officinale)*. Tufted perennial plant with taproot, underground stem, lobed leaves forming a basal rosette; and hollow flower stalk bearing a flower head composed entirely of strapshaped ray flowers. *a.* Achene with tiny curved teeth on the upper margins, long slender beak topped by an umbrellalike tuft of whitish hairs.

## CROWNBEARD, golden crownbeard
SUNFLOWER FAMILY — Compositae

### CROWNBEARD — *Verbesina encelioides* Gray var. *exauriculata* Rob. & Greenm.

DESCRIPTION — A grayish-green branching annual with a rank odor, from a taproot, 1 to 4 feet high reproducing only by seed. The irregularly toothed leaves are densely gray-hairy beneath, green above. The lower leaves are opposite, narrowly triangular, 2 to 8 inches long including the long stalks, which do not have a pair of basal lobes. The other leaves are alternate, lanceshaped, shorter, often with a pair of narrow leaflike lobes, one on each side of the stalk where it joins the stem. The flower heads are on long stalks, 1 to 1¾ inch across, including the 10 to 15 rays. The ray flowers are yellow-orange, deeply 3-notched at the apex, and ⅓ to ½ inch long. The central or disk flowers are also yellow and tubular. The achenes are flattened, covered with fine soft hairs, gray-brown, about ¼ inch long, broadly winged along each margin, with somewhat corklike wings. There are 2 short awns, one on each side at the tip of the achene.

DISTRIBUTION — A weed of disturbed soil, crownbeard is introduced into the southwestern United States from the Old World. Primarily a weed of roadsides and waste places, forming showy masses along the highways and byways throughout Arizona, particularly after the onset of the summer rains; up to 7,000 feet elevation, but usually much lower; flowering April to December, but mostly in late summer.

322

Fig. 156. Crownbeard *(Verbesina encelioides* var. *exauriculata).* Plant with oval and lanceshaped leaves, also flower heads with showy outer (ray) flowers.
*a.* Achene with winged margins and two short awns.

# ANNUAL GOLDENEYE
SUNFLOWER FAMILY — Compositae

## ANNUAL GOLDENEYE — *Viguiera annua* (Jones) Blake

DESCRIPTION — Annual goldeneye is a bushy, sparingly leafy annual, 1 to 3½ feet high, reproducing only by seed. The slender stems are reddish brown and many branched, with the leaves spaced far apart. The narrow leaves are mostly opposite, 1 to 3 inches long, and only 1/16 to ⅛ inch broad.

There are numerous yellow flower heads, about 1 inch broad, including the 12 bright yellow, outer, petallike ray flowers, and the small central (disk) flowers. The heads are sunflowerlike, and are borne at the tips of all the branches. The slender topshaped achenes are blackish, 4-angled, and about 1/16 inch long. They bear no crown of scales nor bristles of any type.

DISTRIBUTION — A native range weed, annual goldeneye is found on open disturbed soil on ridges, plains, and bottomlands, where it replaces valuable forage plants on run-down ranges. Abundant in pinyon-juniper woodland and desert grasslands throughout the state, except in the western part; 2,500 to 8,000 feet elevation. It is especially prolific in Yavapai, eastern Mohave, Santa Cruz, and Cochise counties, where it may occur in solid yellow stands, beautifying large areas several miles long, particularly in early fall after the summer rains; flowering from May to October, but mostly in September and October.

POISONOUS PROPERTIES — Although cattle losses do occur from eating annual goldeneye, little is known about the actual nature of the poison, nor are the conditions necessary for poisoning understood. Animals may consume large amounts with no harm, and at other times small quantities may be fatal. From the symptoms, either nitrate or cyanide poisoning is suspected. Cattle are the only livestock affected. Poisoning usually occurs in the fall when annual goldeneye growth is at its peak.

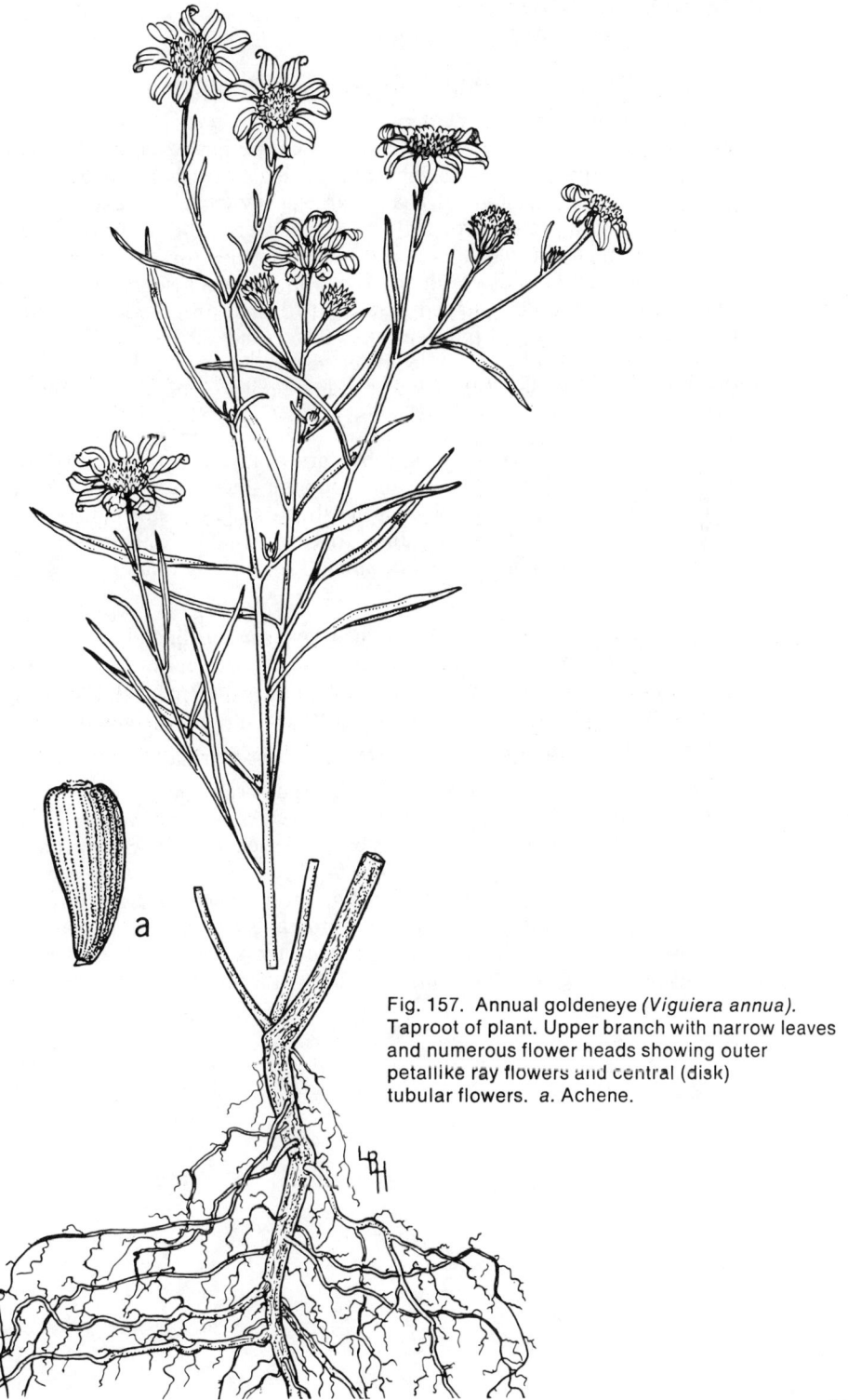

Fig. 157.  Annual goldeneye *(Viguiera annua)*.
Taproot of plant. Upper branch with narrow leaves
and numerous flower heads showing outer
petallike ray flowers and central (disk)
tubular flowers.  *a.* Achene.

a

# COMMON COCKLEBUR

SUNFLOWER FAMILY — Compositae

## COMMON COCKLEBUR — *Xanthium strumarium* L.
### (*X. saccharatum* Wallr.)

DESCRIPTION — Common cocklebur is a coarse bushy annual with stout, usually red-spotted stems, 2 to 3 feet high; reproducing by seed. The large rough glandular green leaves are longstalked, triangular, somewhat lobed, about 2 to 14 inches long, and 1 to 8 inches broad. The short flowering branches arise from the leaf axils along the main stems. The inconspicuous male flowers are grouped into several to many round clusters at the top, with the conspicuous brown female burs at the base. The footballshaped burs, ½ to 1 inch or more long, enclose 2 female flowers, and are covered by about 400 stiff, glandular-hairy spines, ⅛ to ¼ inch long, ending in a hook. The male flowers drop off quickly, but the burs persist, with the 2 blackish achenes. The 2 seeds inside, about ½ inch long, remain fertile for many years.

DISTRIBUTION — Common cocklebur grows in moist flooded soil of roadsides, cultivated fields, pastures, and flats throughout the state, particularly troublesome in wet years around water holes on the sheep ranges of northern Arizona; 100 to 6,000 feet elevation; flowering June to October. The vicious burs form tangled clots in the manes, tails, or wool of animals, often resulting in a lower value of the wool.

POISONOUS PROPERTIES — The seeds and the seedling plants of cockleburs are particularly poisonous to hogs; however, sheep, cattle, horses, and chickens have also been poisoned by eating the seedlings. The seeds are rarely eaten; the seedlings contain the poisonous principle which decreases rapidly as the seedling plant grows.

## SPINY COCKLEBUR — *Xanthium spinosum* L.

DESCRIPTION — Only 2 species of cocklebur occur in Arizona, but they do not resemble one another. Spiny cocklebur has 1 or more stout yellow 3-forked spines, ½ to 1 inch or more long, at the base of each leaf. The long narrow pointed leaves, 1 to 5 inches long, mostly 2- to 5-lobed, are silvery white beneath and dark green above, with the midvein white. The short-spined burs are only ⅓ to ½ inch long. Spiny cocklebur grows in similar situations as common cocklebur, but is not common. Areas where it is troublesome are the Chino, Skull, and Peeples valleys in Yavapai County; also in Santa Cruz and Pima counties.

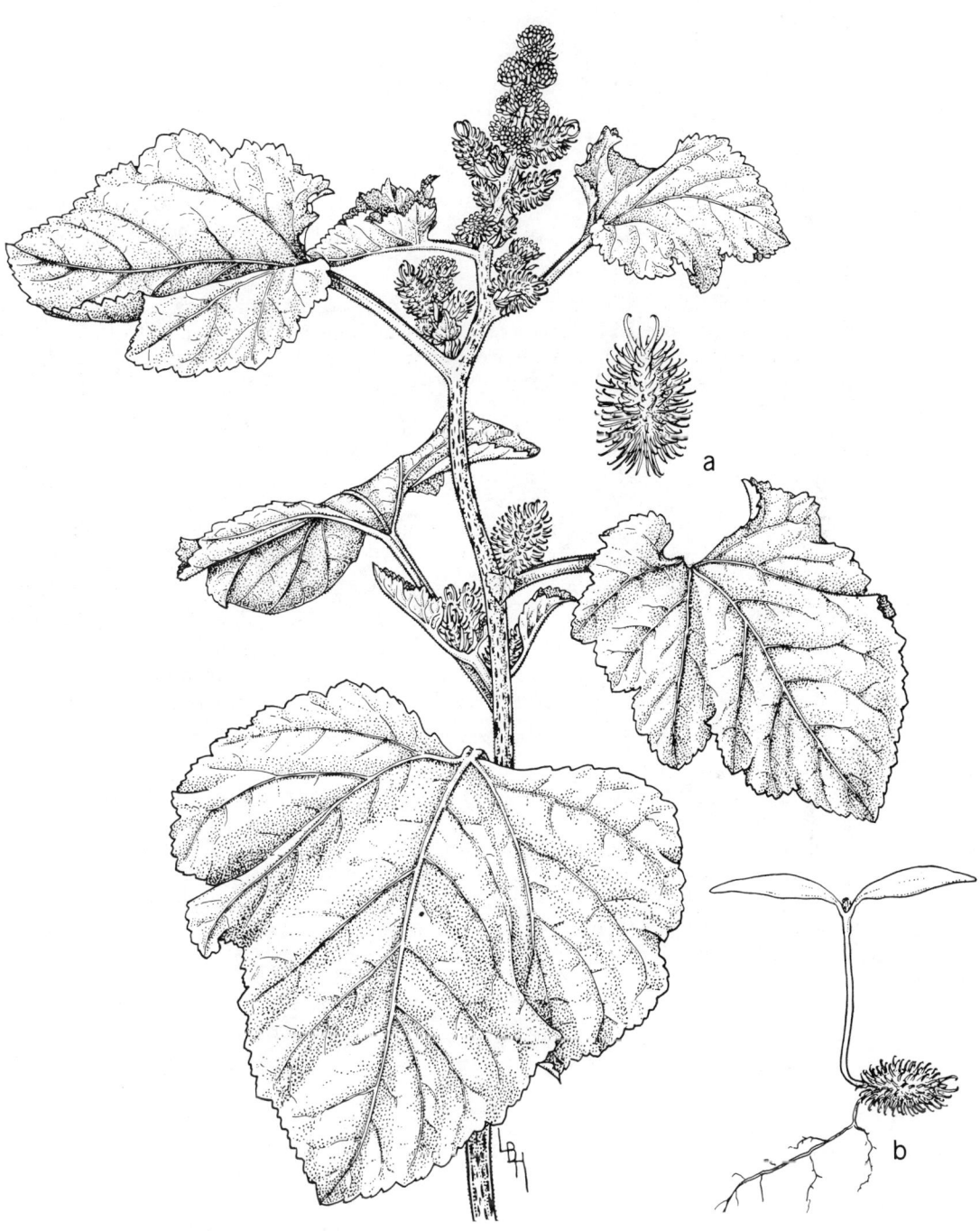

Fig. 158. Common cocklebur *(Xanthium strumarium)*. Leafy branch with large, triangular, longstalked leaves, clusters of male flowers at the top of the stem and the spiny female burs below them. *a.* Spiny female bur enclosing the two achenes, each with one seed. *b.* Seedling emerging from bur.

SECTION 3

# REFERENCES

# BIBLIOGRAPHY

Benson, L., and Darrow, R. A. 1954. *Trees and Shrubs of the Southwestern Deserts.* Tucson: Univ. of Arizona Press and Albuquerque: Univ. of New Mexico Press.

Bohmont, D. W., and Alley, H. P. 1961. *Weeds of Wyoming.* Univ. of Wyoming Agric. Expt. Sta. Bull. 325R.

Correll, D. S., and Johnston, M. C. 1970. *Manual of the Vascular Plants of Texas.* Renner, Texas: Texas Research Foundation.

Frankton, C. 1955. *Weeds of Canada.* Rev. 1963. Ottawa, Canada: Canada Dept. of Agric. publ. 948.

Gilkey, H. M. 1957. *Weeds of the Pacific Northwest.* Publ. Committee and Agric. Expt. Sta. of Oregon State College.

Hamilton, K. C., and Arle, H. F. 1958. *Weeds of Crops in Southern Arizona.* Arizona Agric. Expt. Sta. Bull. 296.

Holmgren, A. H. 1958. *Weeds of Utah.* Utah State Univ. Logan Agric. Expt. Sta. Spec. Rep. 12.

Hotchkiss, N., and Dosier, H. L. 1949. Taxonomy and Distribution of N. Amer. Cattails. *Amer. Midl. Naturalist.* 41: 237–254.

Isley, D. 1960. *Weed Identification and Control in the N. Central States.* Ames: Iowa State University Press.

Kearney, T. H., and Peebles, R. H. 1960. *Arizona Flora.* 2d ed. Berkeley, Calif.: Univ. of Calif. Press.

Kingsbury, J. M. 1964. *Poisonous Plants of the U.S. and Canada.* Englewood Cliffs, N.J.: Prentice-Hall Inc.

Krieger, L. C. C. 1967. *The Mushroom Handbook.* N.Y.: Dover Publ. Inc.

Little, E. L. 1950. *Southwestern Trees.* USDA Agricultural Handbook no. 9. Washington, D.C.

Martin, A. C., and Barkley, W. D. 1961. *Seed Identification Manual.* Berkeley, Calif.: Univ. of Calif. Press.

Nebraska State Dept. of Agric. 1968. *Nebraska Weeds.* Weed and Seed Division. Bull. No. 101-R., Rev. Ed.

Reed, C. F. 1970. *Selected Weeds of the U.S.* Washington, D.C.: U.S. Agric. Research. USDA Agric. Handbook No. 366.

Robbins, W. W.; Bellue, M. D.; and Ball, W. S. 1951. *Weeds of Calif.* Calif. State Dept. of Agric.

Schmutz, E. M.; Freeman, B. N.; and Reed, R. L. 1968. *Livestock-Poisoning Plants of Arizona.* Tucson: Univ. of Ariz. Press.

Thornton, B. J., and Harrington, H. D. n.d. *Weeds of Colorado.* Fort Collins, Colo.: Colo. State Univ. Bull. 514-S.

USDA Agric. Research Center, Beltsville, Md. 1968. *State Noxious-Weed Chart and Supplement.*

WSA Terminology Committee. 1971. Report of the Terminology Committee, Weed Society of America. *Weeds.* 19: 435–476.

# INDEX

**333**

**335**